Contents

DISCARDED

Getting the most from this book 4
About this book 5

Content Guidance

Homeostasis 7
The kidney and excretion 7
Osmoregulation 11
Immunity 13
Coordination and control 23
Coordination in flowering plants 23
Coordination in mammals 29
Ecosystems 41
Populations 41
Communities 47
Ecological energetics 50
Nutrient cycling 57
Human impact on the environment 60

Questions & Answers

Exemplar paper 1 71
Section A 71
Section B 89
Exemplar paper 2 92
Section A 92
Section B 110

Knowledge check answers 113
Index 115

Getting the most from this book

Examiner tips
Advice from the examiner on key points in the text to help you learn and recall unit content, avoid pitfalls, and polish your exam technique in order to boost your grade.

Knowledge check
Rapid-fire questions throughout the Content Guidance section to check your understanding.

Knowledge check answers

1 Turn to the back of the book for the Knowledge check answers.

Summary

Summaries
- Each core topic is rounded off by a bullet-list summary for quick-check reference of what you need to know.

Questions & Answers

Exam-style questions

Examiner comments on the questions
Tips on what you need to do to gain full marks, indicated by the icon $\textcircled{e}$.

Sample student answers
Practise the questions, then look at the student answers that follow each set of questions.

Question 1

Exemplar paper 1

Section A

Question 1

(a) Read through the following passage about phytochromes, and write the most appropriate word(s)/numbers in the blank spaces to complete the account. (3 marks)

The photosensitive pigment phytochrome (P) exists in two interchangeable forms P₆₆₅ and P₇₃₀. P₆₆₅ absorbs _____ coloured light, while P₇₃₀ absorbs _____ coloured light. During the day P_____ is converted to P_____, while in the dark the process is reversed.

(b) The diagram below shows the flowering responses of a short-day plant and a long-day plant when exposed to a specific light regime.

Explain the different flowering responses stated for these two plant types in terms of P₆₆₅ and P₇₃₀.
(3 marks)

Total: 6 marks

$\textcircled{e}$ This is a straightforward question that you might find at the start of the paper. Part (a) tests recall (AO1) but, if you can remember that visible light varies in wavelength between 400nm and 700nm, then you can work it out. Part (b) requires more thought and the ability to express ideas well.

Student A

(a) red ✓, blue ✗, 660 to 730 ✓ a
(b) The LDP flowers as there is enough light to convert the P₆₆₀ to P₇₃₀, ✓ whereas the SDP doesn't flower as the period of time for darkness isn't long enough to convert P₇₃₀ back to P₆₆₀ ✓ b

$\textcircled{e}$ 4/6 marks awarded a While phytochrome is a blue pigment, it absorbs at the red end of the spectrum. P₇₃₀ absorbs far-red light. 2 marks scored. b The student understands the interconversions of the two forms of phytochrome in this light regime and, while not precisely stating so, clearly understands that P₇₃₀ stimulates flowering in LDPs. However, a clear statement that the accumulated P₇₃₀ inhibits flowering in SDPs is lacking. It is important to understand that P₇₃₀ is the active form, whether stimulatory or inhibitory. 2 marks scored.

Unit 1: Physiology and Ecosystems

71

Examiner commentary on sample student answers
Find out how many marks each answer would be awarded in the exam and then read the examiner comments (preceded by the icon $\textcircled{e}$) following each student answer. Annotations that link back to points made in the student answers show exactly how and where marks are gained or lost.

About this book

The aim of this book is to help you prepare for the A2 Unit 1 examination for CCEA Biology.

The **Content Guidance** section contains everything that you should learn to cover the specification content of A2 Unit 1. It should be used as a study aid as you meet each topic, when you prepare for end-of-topic tests, and during your final revision. There are 10 topics and for each there are *examiner tips* and *knowledge checks* in the margins. Answers to the knowledge checks are provided towards the end of the book. Since A2 biology builds on learning at AS, *synoptic links* with AS topics are featured. Where appropriate, there is a list of the *practical work* with which you are expected to be familiar. At the end of each topic, there is a comprehensive, yet succinct, *summary* of the points covered.

The **Questions and Answers** section provides two exemplar papers for you to try. There are answers written by two students, together with examiner comments on their performances and how they might have been improved. These will be particularly useful during your final revision. There is a range of question styles that you will encounter in the A2 Unit 1 exam, and the students' answers and examiner comments should help with your examination technique.

Developing your understanding

It is important that through your A2 course you develop effective study techniques.
- You must not simply read through the content of this book.
- Your understanding will be better if you are *active* in your learning. For example, you can take the information given in this book and present it in different ways:
  - a series of bullet points to summarise the key points
  - an annotated diagram to show structure and function, for example an annotated diagram of a kidney nephron noting what happens in each region with respect to filtration and reabsorption
  - an annotated graph, for example showing the phases of population growth, with notes explaining what is happening in each phase
  - a spider diagram, for example one on the nerve impulse would include reference to neurone structure, the resting potential, the action potential and its propagation as an impulse, and the factors that increase transmission speed
- Compile a glossary of terms for each topic. Key terms are shown in **bold** (with a few in the margin). For each you should provide a definition. This will develop your understanding of the language used in biology and help you in those question parts where *quality of written communication* is being assessed.
- Write essays on different topics. For example, an essay on the immune response will test your understanding of the entire topic and give you practice for the Section B question.
- *Think* about the information in this book so you are able to *apply* your understanding in unfamiliar situations. Ultimately you will need to be able to deal with questions that set the topic in a new context.

- Ensure that you are familiar with the practical skills expected in this unit so that you can answer questions on *How Science Works*.
- Use past questions and other exercises to develop all the skills that examiners must test.
- Use the topic summaries to check that you have covered all the material that you need to know and as a brief survey of each topic.

The A2 Unit I paper

The A2 Unit 1 examination lasts 2 hours and is worth 90 marks. There are two sections. In Section A all the questions are structured, though there is a variety of question styles. In Section B there is a single question, which may be presented in several parts, and which should be answered in continuous prose.

Be aware that the examiners must test a range of skills. These are called assessment objectives and are described in the CCEA biology specification. At A2, examiners must design the paper to show the following balance of marks:

Assessment Objective	Description	Marks
AO1	Knowledge and understanding	32
AO2	Application of knowledge and understanding	45
AO3	*How Science Works*	13

AO1 is mostly found at the start of the paper, in the initial parts of questions and in Section B. However, the emphasis in the A2 Unit 1 paper is on showing your ability to apply your understanding (AO2), and questions will often be presented in novel contexts.

AO3 may be assessed by questions that ask you to formulate a hypothesis, design an appropriate experiment, evaluate the methodology used in an experiment and analyse results to work out whether a hypothesis has been supported or disproved. You may also be asked to consider the ethical implications of the way research is carried out.

In A2 papers you should also expect that some question parts will build on your learning at AS. This is *synoptic assessment*. You are advised to review your work at AS in preparation for synoptic assessment. Some of the knowledge checks test synoptic links.

You are expected to use good English and accurate scientific terminology in all your answers. Quality of written communication is assessed throughout the paper and is specifically awarded a maximum of 2 marks in Section B.

Examiner tip
The CCEA biology specification is available from www.ccea.org.uk. You will also be able to access past papers and mark schemes.

Content Guidance

Homeostasis

The kidneys and excretion

The kidneys regulate the internal environment by constantly adjusting the composition of the blood. They are the organs of excretion and osmoregulation.

Excretion is the removal from the body of the toxic waste products of metabolic processes. In mammals, carbon dioxide, produced during respiration, is excreted from the lungs. The kidneys excrete nitrogen-containing compounds, mostly **urea**, produced during the breakdown of excess amino acids and nucleic acids in the liver. The kidneys also excrete a little creatinine, a waste product produced from the degradation of creatine phosphate (a molecule of major importance in ATP generation) in muscles.

Homeostasis is the maintenance of steady states within the body. The kidneys have a homeostatic function in regulating the water content of the blood. The kidneys control the water potential of body fluids (**osmoregulation**) under the influence of antidiuretic hormone.

The structure of the urinary system

The components of the urinary system of a mammal are shown in Figure 1.

Figure 1 The urinary system of a mammal

It is important to distinguish the terms *excretion* and *egestion*. **Excretion** is the removal of waste products of metabolism from the body. **Egestion** is the removal from the body of waste material, such as undigested food, which has not been part of body metabolism, i.e. faeces are not excreted; they are egested.

Examiner tip
The examiners often begin a question by asking students to define a term. As a part of your revision, construct a list of definitions for all the terms shown in bold within this chapter. This is an exercise that you should also complete for each of the following chapters.

An indication of the importance of the kidneys in regulating the composition of the blood is the fact that they receive approximately 25% of the cardiac output (via the aorta and renal arteries).

Inside a kidney there are two layers — an outer **cortex** and inner **medulla** — surrounding a central cavity, the **pelvis**. The medulla is sub-divided into a number of **pyramids** whose apices protrude into the pelvis. A kidney contains over one million microscopic tubules called **nephrons**, each of which has a rich blood supply. The positioning of nephrons relative to other regions of the kidney is shown in Figure 2a. The overall structure of a nephron and its blood supply is shown in Figure 2b.

Examiner tip
You may be expected to interpret the histology of the kidney. Make sure that you use the information provided about the region of the kidney from which the section is taken. For example, if it is from the cortex you might expect to see the glomerulus and Bowman's capsule or, alternatively, sections of the convoluted tubules.

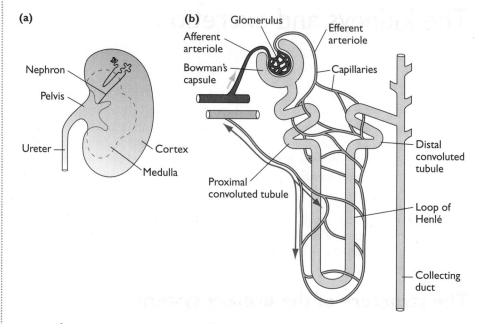

Figure 2 (a) Main regions of a kidney; (b) the nephron and its blood supply

The nephron is the functional unit of the kidney. Each nephron consists of a cup-shaped **Bowman's capsule**, plus a tube that has three distinct regions: the **proximal (first) convoluted tubule**, the **loop of Henlé**, and the **distal (second) convoluted tubule**. Arterial blood enters each capsule through an **afferent arteriole** which branches to form a capillary network called the **glomerulus**; blood leaves the glomerulus through an **efferent arteriole** which then branches to form a further capillary network (the vasa recta system) around the main body of the nephron. A number of nephrons join to form a **collecting duct** which transfers the fluid towards the pelvis.

The production of urine

Urine is produced in two stages involving quite different processes:

- **Ultrafiltration**, in which plasma in the glomerulus is filtered into the Bowman's capsule; only substances below a certain size are filtered and so the filtrate contains useful molecules as well as toxic ones.
- **Reabsorption** of useful substances back into the blood occurs as the filtrate passes along the nephron and collecting duct; only at the point where the collecting duct joins with the pelvis can the fluid be called 'urine'.

Knowledge check 1
Suggest why nephrons have tubules with convolutions.

Ultrafiltration

The driving force: The blood entering the glomerulus is under high pressure. This **high hydrostatic pressure** occurs because:
- the renal arteries are wide, short and relatively close to the heart
- the efferent arteriole is smaller than the afferent arteriole (carrying blood to the glomerulus), which creates a bottleneck

It is this high hydrostatic pressure which effectively causes fluid to filter from the glomerular plasma as filtrate in the capsule. Ultrafiltration is so efficient that 15–20% of the water and solutes are removed from the plasma that flows through the glomeruli.

The filter: There are only three layers separating plasma from filtrate. These are: the capillary endothelium; the basement membrane on which the capillary cells lie; and the inner layer of the Bowman's capsule. Two of these layers are especially porous:
- the endothelium of the capillaries in the glomerulus, which consists of a single layer of squamous (thin) cells with pores between them
- the inner wall of the Bowman's capsule consists of podocytes, with foot-like processes which surround the capillaries but which have spacious gaps between them called filtration slits

The *effective filter* is the **basement membrane** of the glomerular capillaries. This extracellular membrane lies on the outer side of the capillary endothelium. These layers are shown in Figure 3 (not drawn to scale).

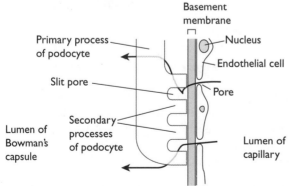

Figure 3 The structure of the filter

Knowledge check 2

What creates the high hydrostatic pressure in the glomerulus?

Knowledge check 3

Explain how the structure of (a) the endothelial cell of the glomerular capillaries, (b) the basement membrane and (c) the podocytes influences their roles in ultrafiltration.

Knowledge check 4

What is the concentration of glucose in the glomerular filtrate relative to its concentration in the blood plasma?

The composition of the filtrate: normally only molecules with a relative molecular mass of less than 68 000 can pass through the basement membrane. All constituents of the blood plasma other than plasma proteins — except for the smallest of these — are able to pass through. The filtrate consists mainly of inorganic ions, glucose, amino acids, urea and other toxic molecules, all dissolved in water. Clearly, water and useful substances must be reabsorbed.

Reabsorption

As the filtrate flows through the **proximal convoluted tubule**, 80% of the water is reabsorbed by osmosis into adjacent blood capillaries. As a consequence, ions follow — partly by diffusion, partly by active transport. All the glucose and amino acids pass back into the blood by active transport. Small proteins, which may have been filtered, are reabsorbed by pinocytosis. By the end of the proximal convoluted tubule, the filtrate is isotonic with the plasma (i.e. it has the same concentration of solutes).

Knowledge check 5

Explain why reabsorption from the nephron must be selective, i.e. some molecules are reabsorbed more readily than others.

The cuboidal epithelial cells, which line the tubule walls, have numerous **microvilli** on the side in contact with the filtrate, and **infoldings** of the cell surface membrane on the side next to the blood capillaries. These adaptations greatly increase the surface area available for reabsorptive processes. Furthermore, the cells have many **mitochondria** located near the infoldings and these supply the extra ATP needed for active transport (see Figure 4).

Examiner tip

You may be asked about adaptations for reabsorption from the proximal convoluted tubule. Significant features that ensure efficiency include: microvilli increasing surface area; infoldings of membranes; abundance of mitochondria; close association of many capillaries; thin squamous epithelium of capillaries; flow of filtrate and blood maintaining concentration gradients.

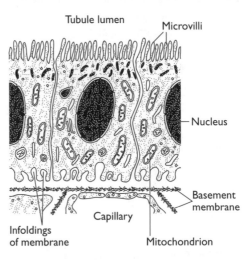

Figure 4 A proximal tubule cell

The **loop of Henlé** operates (through a combination of water and ion movement) to make the tissues deeper in the medulla more concentrated with ions. In other words, this process creates an increasingly negative solute potential through the medulla tissue of the kidney. This is significant in the further reabsorption of water from the collecting ducts (see Osmoregulation, below).

In the **distal convoluted tubule** the ionic composition and the pH of the blood are adjusted. It is also here that toxic substances, such as creatinine, are secreted into the filtrate.

Osmoregulation

The collecting ducts are the place where the water content of the blood (and therefore of the whole body) is regulated. The permeability to water of the walls of the second convoluted tubule and collecting duct is increased by **antidiuretic hormone** (ADH). ADH is *produced* by the **hypothalamus** but secreted into the **posterior lobe of the pituitary body**, where it is stored. A rise in blood concentration (i.e. when the water potential of the blood becomes more negative) is detected by **osmoreceptors** in the hypothalamus; these receptors send impulses to the posterior lobe of the pituitary gland. As a result, this lobe of the pituitary gland releases *more ADH* into the blood which *increases the permeability* to water of the second convoluted tubule and collecting duct. In fact, water moves through channel proteins (aquaporins) which open, under the influence of ADH, to let water through. More water passes to the medulla and a more concentrated (hypertonic) urine is produced.

A fall in blood concentration (i.e. the water potential of the blood becomes less negative) *inhibits the release of ADH*. As a result, the walls of the second convoluted tubule and collecting duct become *impermeable to water*, less water is reabsorbed and a less concentrated (hypotonic) urine is produced.

Osmoregulation involves **negative feedback**: it is *feedback*, since a change in the water potential of the blood (detected by the osmoreceptors and determining the release of ADH) will ultimately lead to a further change in the water potential of the blood; it is *negative*, since an increase in water potential (e.g. blood diluted by drinking) will later result in a decrease in water potential (see Figure 5).

Knowledge check 6

Name (a) the structure responsible for the detection of changes in the water potential of blood and (b) the gland responsible for the release of ADH.

Knowledge check 7

Is a dehydrated person likely to produce more or less ADH?

Knowledge check 8

Suggest why beavers have short loops of Henlé.

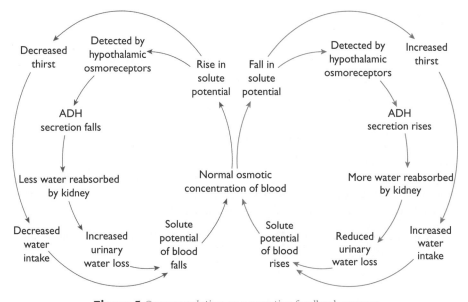

Figure 5 Osmoregulation as a negative feedback process

Synoptic links

Understanding some aspects of the functioning of the kidney relies on how well you understand concepts and facts that you learned at AS. For example, to understand ultrafiltration and reabsorption you need to know that:

- water moves from a region of high water potential to a region of lower water potential
- water potential has two components: the solute and pressure potentials
- if water moves out of a region, the water potential becomes more negative (solutes are more concentrated)
- if salt moves into a region, the water potential is reduced (becomes more negative)

Calculation of the net filtration pressure between glomerulus and capsule

Overall, the water potential of the glomerular plasma exceeds the water potential of the filtrate in the capsule. This is due to the large pressure potential (high hydrostatic pressure) within the glomerulus. This is, in part, opposed by a more negative solute potential within the plasma (in which the retained proteins act as solutes) than in the filtrate, while there is some resistance to further filtration due to back pressure of the filtrate in the nephron.

The net filtration force is the difference in water potential either side of the filter, i.e. water potential of glomerular plasma minus water potential of filtrate in the capsule. Net filtration force is calculated as follows:

$$\Psi_{plasma} = \Psi_s \text{ (due to proteins)} + \Psi_p \text{ (hydrostatic pressure)}$$

$$= -3.3\,kPa + 6\,kPa$$

$$= 2.7\,kPa$$

$$\Psi_{filtrate} = \Psi_s \text{ (no proteins present)} + \Psi_p \text{ (hydrostatic pressure)}$$

$$= 0 + 1.3\,kPa$$

$$= 1.3\,Pa$$

Therefore, the net filtration force $= \Psi_{plasma} - \Psi_{filtrate} = 1.4\,Pa$

You should also know:

- cell ultrastructure — particularly the role of mitochondria and microvilli
- how substances can move across the cell-surface membrane: by diffusion, facilitated diffusion, active transport and pinocytosis

These topics are covered in the AS Unit 1 Student Unit Guide, to which you should refer for revision.

Examiner tip

Note that the water potential of the plasma and filtrate are positive (cell water potentials are generally negative — see AS Unit 1). This is simply because the plasma is pumped (by the heart) through the kidney, creating a pressure potential of great magnitude.

- Excess amino acids in the diet are deaminated (in the liver), producing urea, which enters the blood circulatory system to be transported to the kidneys.

- The kidney is composed of about a million small tubules, called nephrons, organised within distinct layers — the outer cortex and the inner medulla.

- Each nephron is supplied with blood vessels, which form a knot of capillaries, the glomerulus, enclosed by the cup-shaped Bowman's capsule, and which surround all other parts of the nephron.

- Each nephron produces urine as follows:

  - Ultrafiltration takes place from the glomerulus into the Bowman's capsule. Small molecules are filtered, by the basement membrane of the glomerulus endothelium, under pressure. The glomerular filtrate is hypotonic to the blood as it lacks plasma proteins.

  - In the proximal convoluted tubule, glucose, amino acids and some ions are actively transported back into the blood during selective reabsorption. Water follows by osmosis and the filtrate becomes isotonic to the blood.

  - The loop of Henlé creates hypertonic conditions in the medulla, giving a decreasing water potential gradient from the boundary with the cortex to the inner medulla. This enables water to be withdrawn from the collecting ducts if they are permeable to water.

  - Before the filtrate passes down the collecting duct, it travels through the distal convoluted tubule where further ions are actively reabsorbed.

- Antidiuretic hormone (ADH) makes the wall of the distal convoluted tubule and collecting ducts permeable to water by causing water channel proteins, called aquaporins, to open.

- ADH is produced in the hypothalamus but stored in the posterior lobe of the pituitary gland from where it is secreted into the blood when hypothalamic osmoreceptors detect that the blood water potential has fallen below the norm. ADH enables water to be reabsorbed through the tubule walls so that hypertonic urine is produced.

- If the water potential of the blood rises, then ADH release is reduced and the impermeability of the tubule walls prevents the reabsorption of water. More water passes out in the dilute urine and the water potential of the blood falls.

Immunity

The conditions in the body provide an ideal environment for the growth of a vast array of micro-organisms. These include viruses, bacteria, fungi and protoctists, which are present in the air we breathe, in the water and food we consume and on the objects we touch. However, relatively few of these micro-organisms are disease-causing, or **pathogenic**.

Examples of some infectious or communicable diseases, the micro-organisms that cause them, the tissues infected and the symptoms of the diseases are provided in Table 1.

Knowledge check 9

Describe the differences between a parasite and a pathogen.

Table 1 Examples of infectious diseases caused by a variety of pathogenic micro-organisms

Disease	Micro-organism	Tissue infected	Symptoms
Measles	An RNA virus of the genus *Morbillivirus*	Gas exchange system	Fever and a characteristic rash
Poliomyelitis ('polio')	An RNA virus of the genus *Enterovirus*	Nervous tissue	In the most severe cases paralysis occurs
Acquired immune deficiency syndrome (AIDS)	Human immunodeficiency virus — a retrovirus of the genus *Lentivirus*	Helper T-cells of the immune system (specifically CD4$^+$ T-cells) — see later	Occurrence of diseases normally controlled by the immune system that HIV damages

Table 1 (*continued*)

Disease	Micro-organism	Tissue infected	Symptoms
Tetanus ('lockjaw')	The anaerobic bacterium *Clostridium tetani*	The bacteria produce a toxin that passes to all parts of the body	Prolonged contraction of skeletal muscle fibres
Tuberculosis ('TB')	The bacterium *Mycobacterium tuberculosis*	The lungs	Coughing, shortness of breath, fever and extreme fatigue
Athlete's foot	A fungus of the genus *Trichophyton*	The upper layer of the skin (when moist and warm)	Itching of the skin, which may crack and peel
Malaria	A protoctist of the genus *Plasmodium*, transmitted by the *Anopheles* mosquito	Red blood cells and liver tissue (where *Plasmodium* can remain dormant)	Anaemia, fever, chills, nausea and, in severe cases, coma and death

Examiner tip

The infections listed in Table 1 are only provided so that you may be aware of the causes and effects of some examples. You do not have to learn them.

For a micro-organism to be pathogenic it must (1) enter the host, (2) colonise tissues of the host, (3) evade the host's defences and (4) cause damage to the host tissues. If a pathogen gets into the host and colonises its tissues, an **infection** results. **Disease** occurs when an infection leads to recognisable symptoms in the host. The disease may cause damage to host tissues directly (e.g. viruses cause cells to break down) or through the production of toxins (e.g. as many bacteria do).

The body has three lines of defence to resist pathogens:
- the first is to prevent the entry of pathogens
- if this fails, the second is for phagocytes to gather at the site of infection where they ingest pathogens
- if this fails, the third is for the body to target that particular pathogen, in a specific immune response

Knowledge check 10

Suggest how bacteria living in your intestines can help to prevent infection of the gut by pathogenic bacteria.

Barriers to pathogen entry

The barriers to the entry of pathogens are:
- the **skin**, the outer layer of which consists of layers of dead cells filled with a tough protein (keratin) and covered in an oily secretion (sebum)
- **tears** and **saliva**, which contain the enzyme **lysozyme** capable of digesting the cell walls of many bacteria
- **mucus**, secreted by cells lining the respiratory tract, which traps micro-organisms and prevents them penetrating the underlying membranes, while cilia normally sweep this mucus up and out
- **acid** (specifically hydrochloric acid), secreted in the stomach, which kills most of the bacteria entering in the water and food consumed

Nevertheless, pathogens can still enter the body, most readily through broken skin (e.g. as the result of a cut) or via the large absorptive surfaces in the lungs and the intestines.

The activity of phagocytes

If a pathogen does get into the body, changes occur in the infected area to cause the capillaries to become leaky. As plasma escapes, the area becomes swollen (in what is known as an inflammatory response) but, more importantly, phagocytic white blood cells readily squeeze through the leaky capillary walls and accumulate in the area. These phagocytes include:

- **polymorphs**, which are the most common phagocytes and so the first to arrive
- **macrophages**, which develop from the monocytes in the blood, and are larger and longer-lived

Both types of phagocyte engulf the bacteria and the debris from damaged cells. Ingested material is enclosed within a vacuole; then, lysosomes fuse with the vacuole, releasing hydrolytic enzymes which destroy the bacteria. The process is shown in Figure 6.

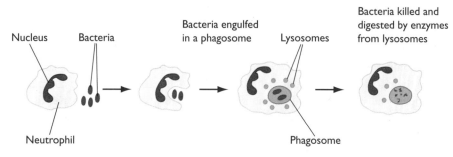

Figure 6 The activity of a phagocyte in engulfing and destroying bacteria

The specific immune response

The **immune response** is a specific response to the detection of pathogens in the body. It involves **lymphocytes**, another type of white blood cell. The pathogens carry molecules on their outer surfaces which are recognised by the body as being **foreign** or **'non-self'**. These molecules may consist of protein, carbohydrate or glycoprotein. They initiate the immune response, and are called **antigens**. Different pathogens have different antigens. Lymphocytes all look the same; however, they differ biochemically. For each antigen that occurs on the surface of a pathogen there is a lymphocyte that carries a special **protein receptor** on its cell-surface membrane which is complementary in shape. It is the complementary nature of the receptor and antigen (similar to the lock-and-key model of enzyme action) that ensures a specific response.

During early development, many millions of different lymphocytes are produced from the stem cells in the bone marrow. Each carries a specific membrane receptor that allows it to respond to a different non-self antigen if this should be encountered in the future. This gives the immune system the ability to respond to any type of pathogen that enters the body. There are no lymphocytes set up to respond to any molecules on the cell-surface membranes of that individual's own cells ('self' antigens).

In fact, there are two sets of lymphocytes: **B lymphocytes** and **T lymphocytes** (also called **B-cells** and **T-cells** respectively; see Table 2). Both types are produced in the bone marrow, but they differ in terms of where they mature and the nature of the immune response produced.

Table 2 A comparison of B and T lymphocytes

Type of lymphocyte	Maturation	Type of immune response	Nature of immune response
B lymphocytes (B-cells)	Continue maturing in the **B**one marrow	Antibody-mediated (humeral) immunity	Secrete antibodies which counter the antigen-carrying pathogens
T lymphocytes (T-cells)	Mature in the **T**hymus	Cell-mediated immunity	Attack infected cells with the antigen presented on the surface

Both B and T lymphocytes migrate to lymphoid tissue throughout the body to await possible activation.

Activation of a lymphocyte involves it coming into contact with an antigen that its receptor recognises. Antigens are present on the pathogen, while infected cells display antigens on their surface (which is how T lymphocytes detect infected cells). There is a delay between entry of the pathogen and activation of the appropriate lymphocyte. There are a number of reasons for this.

There is only one or a few of the 'correct' B and T lymphocytes available, so it takes time for them to come into contact with the antigen. Once they have made contact with the antigen, the lymphocytes become **sensitised**. During this time, a specific gene is activated, mRNA synthesised and subsequently proteins produced. These proteins are either antibodies in B-cells, or membrane receptors in T-cells. This process also takes time. The sensitised lymphocytes then divide by mitosis many times — they are **cloned**. Again this takes time. Overall, there is a delay of about 4 days, between contact with the antigen and the cloning of the required lymphocytes, during which time the person suffers from the disease caused by the pathogen.

Antibody-mediated immunity

Most of the cloned B-cells develop into **plasma cells** which synthesise and secrete large amounts of the antibody. Their activity is intense — they may produce several thousand antibodies per second — and so they are short-lived. After several weeks their numbers decrease but the antibodies remain in the blood for some time. Eventually, however, the concentration of antibodies decreases too.

Antibodies are globular proteins called immunoglobulins. An antibody has a binding site with a shape complementary to a particular antigen (see Figure 7) and forms an antigen–antibody complex.

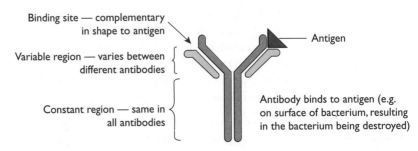

Binding site — complementary in shape to antigen

Variable region — varies between different antibodies

Constant region — same in all antibodies

Antigen

Antibody binds to antigen (e.g. on surface of bacterium, resulting in the bacterium being destroyed)

Figure 7 The structure of an antibody (binding to an antigen)

Antibodies function in a variety of ways:

- Some may neutralise toxins produced by bacteria — these antibodies are called **antitoxins**.
- Some antibodies clump or agglutinate bacteria before the latter are engulfed by phagocytic cells — these antibodies are **agglutinins**.
- Some antibodies attach to viruses, preventing them from entering and infecting host cells.
- Some antibodies destroy bacterial cell walls, causing lysis (the cells burst).
- Some antibodies attach to bacteria, enabling phagocytic cells to identify them.

Some of the cloned B-cells remain in the blood but they are not plasma cells and do not secrete antibodies. They are **memory cells**, and they live for a very long time. If the same antigen is encountered again the memory cells rapidly clone to produce plasma cells which secrete antibodies. This rapid response means that no 'infection' is suffered on the subsequent occasion — the person has become immune to that particular disease.

The way in which antibody-mediated immunity defends against bacteria is illustrated in Figure 8.

1 Bacteria invade body — antigens on bacterial surface

2 Bacterial antigen recognised by the correct B-cell with the complementary receptor

Correct B-cell   Other B-cells with different receptor sites

3 The sensitised B-cell divides by mitosis to produce plasma cells and memory cells

4 Plasma cells produce antibodies, which destroy bacteria

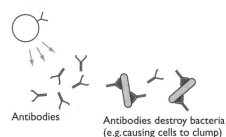

Figure 8 Antibody-mediated immunity

Knowledge check 12

A collision between an appropriate B-cell and its antigen is less likely than the collision between a memory cell and the same antigen. Explain why.

Cell-mediated immunity

Cloned T-cells form a variety of cells:

- **Killer (cytotoxic) T-cells** destroy infected cells directly by attaching to the antigens on the surface of the target (infected) cell and releasing perforin, which punches holes in the cell-surface membrane, resulting in lysis and death of the target cell. Hydrolytic enzymes are also secreted to aid in this destruction.
- **Helper T-cells** promote the activity of other cells: they stimulate B-cells to produce plasma cells and so increase antibody production; they activate killer T-cells; they activate macrophages and so stimulate phagocytosis.

Knowledge check 13

What is the difference between helper T-cells and killer T-cells?

- **Memory T-cells** do not act immediately but multiply very quickly if the antigen appears again, producing an even bigger crop of cloned T-cells, which results in the rapid destruction of any cells that present the antigen.

The way in which *cell-mediated immunity defends against viruses* is illustrated in Figure 9.

1 Host cell infected with viruses presents viral antigens on its surface membrane

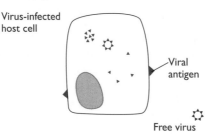

2 The viral antigen is recognised by the correct T-cell with the complementary receptor

Knowledge check 14

HIV (human immunodeficiency virus) invades and destroys helper T-cells and may lead to AIDS (acquired immune deficiency syndrome). Explain why people with full-blown AIDS suffer infections that would normally be held in check (called *opportunistic infections*).

3 The sensitised T-cell divides by mitosis to produce different types of T-cell

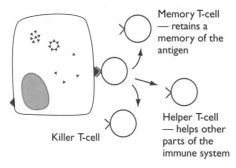

4 Killer T-cells recognise infected cells and destroy them before viruses reproduce

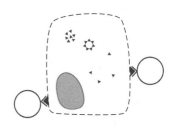

Figure 9 Cell-mediated immunity

The *T and B systems complement each other* in their action. For example, plasma (B) cell antibodies destroy 'free' pathogens but cannot deal with pathogens that are inside host cells; killer T-cells destroy those infected cells. Many bacteria damage tissue without entering cells, and so these, and any free viruses in the plasma, are dealt with by the B system; all viruses and a few bacteria invade host cells, and such infected cells are dealt with by the T system. T-helper cells enhance other parts of the immune system.

Killer T-cells destroy not only infected cells, but also:
- **cancer (tumour) cells** within the host, since these present abnormal antigens on their cell-surface membranes
- **cells of transplanted tissue**, since cells from another person — unless that person is an identical twin — will have antigens which are distinct

Active and passive immunity

Passive immunity occurs when an individual receives antibodies from another source. This can happen naturally or artificially.

- **Natural passive immunity** occurs when antibodies pass naturally from mother to baby across the placenta and in mothers' milk (colostrum).
- **Artificial passive immunity** occurs when antibodies which have been made in one individual are injected into another person, as a serum.

Passive immunity is only temporary since antibodies naturally degrade in the body and the recipient has no capacity to make more. However, passive immunity will provide a baby with protection until the child develops its own immune system. Passive immunity is the only way to save someone bitten by a fatally venomous snake or spider — the victim is injected with a serum containing antibodies (antitoxins) against the toxin. Figure 10 shows what happens to the level of antibodies in an individual when these are received passively.

Figure 10 Changes in the concentration of antibodies during passive immunity

Active immunity occurs when an antigen enters the body and stimulates the body's immune system to produce antibodies (and killer T-cells) and, most importantly, memory cells. Since memory cells can last a lifetime, *active immunity provides long-term immunity*. Active immunity happens naturally when you get a disease, or it may be artificially induced.

Natural active immunity happens when a person is infected. On the first occasion, the person suffers the disease while B- and T-cells are activated and cloned — there is a time delay until antibodies are produced. This initial period is called the **primary response**. As the antibodies destroy the antigens, fewer and fewer B-cells are made and the concentration of antibodies falls again. If the same person is later infected with the same antigen (pathogen), the response is more rapid and a greater quantity of antibodies is produced. This **secondary response** is due to the action of memory cells:

- The response is rapid because the memory cells have already been activated and cloning takes place immediately.
- Greater quantities of antibodies are produced as there are many memory cells producing plasma cells (in comparison to only one or a few 'correct' B-cells initially).

As a result, a person will not (normally) suffer the same infection twice. The primary and secondary responses are shown in Figure 11.

> **Knowledge check 15**
>
> Explain why bottle-fed babies are more likely to have infections than breast-fed babies.

> **Knowledge check 16**
>
> For B and T lymphocytes, give two differences and two similarities.

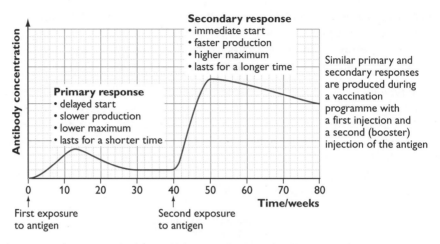

Figure 11 Changes in the concentration of antibodies within the primary and secondary responses during active immunity

Knowledge check 17

How would the blood of someone who was immune to a disease be different from the blood of someone who was not immune to that disease?

Artificial active immunity involves vaccination (or immunisation). **Vaccination** was developed because some diseases are so serious that you really would not want to suffer their consequences even once. It involves injecting a person (or animal) with antigenic material that has been rendered harmless. However, the immune system treats the antigenic material as a real pathogen. As a result, the immune system manufactures antibodies and memory cells. The memory cells provide the long-term immunity. For some diseases, a vaccination programme involves a second injection of the vaccine, or **booster**. This further heightens the immune system — the level of antibodies produced by the two injections is similar to that produced as a result of two bouts of infection, as illustrated in Figure 11.

The different types of immunity are summarised in Table 3.

Table 3 Summary of different types of immunity

	Active: body receives antigens	**Passive: body receives antibodies**
Natural	Natural active: fighting infection	Natural passive: from mother via placenta or milk (colostrum)
Artificial	Artificial active: vaccination, involving injection of antigens or attenuated (weakened) pathogen	Artificial passive: injection of antibodies, e.g. antitoxins

Organ transplant

Proteins (and glycoproteins) on the cell-surface membrane act as labels or markers and, in any one individual, are treated by the immune system as 'self'. However, since people differ genetically (except in the case of identical twins), different individuals will have different molecules on the surfaces of their cells. This presents a major problem when an organ or some tissue is transplanted from one individual to another. The recipient's body will tend to 'reject' the donated organ due to the activity of killer T-cells.

Successful organ transplants rely on the following:

- **Tissue typing**, where the compatibility of donor and recipient cell-surface molecules (markers) is first determined, then donor tissue is used for which there is an optimal match (i.e. most markers are the same). Tissue matching is more likely to occur between relatives (especially close ones) than between non-relatives.

- **Use of X-rays** to irradiate bone marrow and lymph tissues, so as to inhibit the production of lymphocytes and therefore slow down rejection. However, unpleasant side-effects occur and the patient is at increased risk of infection while the treatment is going on.
- **Immunosuppression** through the use of drugs, some of which inhibit DNA replication, cell division and the cloning of lymphocytes, and so delay the rejection of the graft. Again, problems may develop, including an increased susceptibility to infection.

Blood transfusion

Blood can be classified into types (groups) according to the *different markers on the surface of red blood cells*. These markers act as antigens and affect the ability of red blood cells to provoke an immune response. Since the markers fall into distinct types they represent blood group polymorphisms (i.e. cases with many distinct forms).

The **ABO blood system** is the most important blood-type system, because of the presence of anti-a and anti-b antibodies in people who lack the corresponding antigens from birth (see Table 4).

Table 4 The antigens and antibodies of the ABO blood group system

Blood group	Antigens on red blood cells	Antibodies in plasma
A	A	anti-b
B	B	anti-a
AB	both A and B	neither anti-a nor anti-b
O	neither A nor B	both anti-a and anti-b

Anti-a and anti-b are agglutinins, i.e. if they encounter red blood cells with the corresponding antigen they will cause them to stick together or agglutinate (clump; see Figure 12).

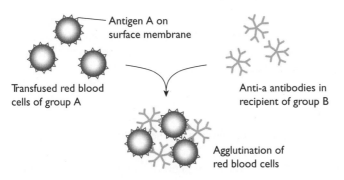

Antigen A on surface membrane

Transfused red blood cells of group A

Anti-a antibodies in recipient of group B

Agglutination of red blood cells

Figure 12 Donor blood group A will be agglutinated by anti-a antibody in the plasma of a recipient with blood group B

This effect is of great importance in **blood transfusion**: if donated blood were to agglutinate in the recipient's blood system, the clumps (note: not *clots*) of blood formed would block capillary networks and cause organ failure and death. The donated blood does not have to have the same blood type as the recipient. However, it is important that the donated blood does not have antigens that would be recognised by any of the antibodies present in the recipient. The *'safe' and 'unsafe' transfusions* are shown in Table 5.

Table 5 Safe ✔ and unsafe ✗ transfusions: the recipient must not possess antibodies that will recognise antigens on the red blood cells of the donor

Blood group of transfused blood	Blood group of recipient			
	A	B	AB	O
A	✔	✗	✔	✗
B	✗	✔	✔	✗
AB	✗	✗	✔	✗
O	✔	✔	✔	✔

Blood of group O lacks the A and B antigens and is safe to donate to all other types, so people of this group are called **universal donors**; blood of group AB lacks both antibodies, so people with this type can safely receive blood of any blood type and are called **universal recipients**.

Other ways of grouping or classifying blood exist. The **Rhesus system** has two types, positive (Rh+ve) or negative (Rh–ve), according to whether or not a marker protein is present. People who are Rh+ve will, of course, not produce the Rh antibody (anti-D). However, people who are Rh–ve, whose blood has been contaminated by red blood cells with the Rh antigen, either during a blood transfusion or during labour, will produce anti-D. With respect to transfusions, an Rh–ve person can only safely receive Rh–ve blood, while an Rh+ve person can receive both types. However, there are *potential complications if a pregnant woman is Rh–ve and the baby is Rh+ve* (fetal blood cells may be forced across the placenta as a result of severe uterine contractions during delivery). Generally, the first Rh+ve baby is born before the Rh–ve mother produces the anti-D. However, any subsequent Rh+ve baby may receive anti-D from the mother, with the result that the baby's blood agglutinates. One possible way to eliminate the problem is to give the fetus several blood transfusions as it develops in the uterus. An alternative solution is to inject an Rh–ve mother with anti-D immediately after she has given birth to an Rh+ve baby; this injected anti-D will destroy fetal blood cells in the mother's body before the latter is sensitised to make its own antibodies.

Synoptic links

You need to have an understanding of the following topics, which you should revise:
- the role of endoplasmic reticulum in producing proteins (e.g. antibodies) and of the Golgi body in producing lysosomes (see the AS Unit 1 Student Unit Guide)
- phagocytosis (a form of endocytosis) and the role of lysosomes (see AS Unit 1 Student Unit Guide)
- membrane structure and, especially, the range of molecules on the external surface (see AS Unit 1 Student Unit Guide)
- the structure of viruses and their functioning only inside a host cell (see AS Unit 1 Student Unit Guide)

- blood cells and, especially, the different types of white blood cell (see the AS Unit 2 Student Unit Guide)
- specificity of globular protein structure (see AS Unit 1 Student Unit Guide)

- Resistance to infection by pathogenic organisms is possible because of barriers to entry, phagocytosis (by polymorphs and macrophages) and specific immune responses (antibody-mediated and cell-mediated).
- B lymphocytes are involved in antibody-mediated immunity while T lymphocytes are responsible for cell-mediated immunity.
- Antigens are specific molecules, found on the surface of invading pathogens (e.g. bacteria) or on infected cells, which trigger an immune response.
- In antibody-mediated immunity:
  - an inactive B lymphocyte with complementary receptors on its surface binds to a foreign (non-self) antigen on a microorganism or virus
  - the sensitised B lymphocyte divides to form plasma cells and memory B-cells
  - the plasma cells secrete antibodies that combine with the antigens on the pathogen and neutralise them (e.g. by agglutination)
  - the memory B-cells remain for many years but form plasma cells if the same antigen is detected in the future
- In cell-mediated immunity:
  - a T lymphocyte with complementary receptors binds to a foreign antigen displayed on an infected cell (or cell changed by cancer or cell transplanted from another individual)
  - the sensitised T lymphocyte divides to form helper T-cells, killer T-cells and memory T-cells

  - helper T-cells produce chemicals that stimulate the B lymphocytes, killer T-cells and the phagocytes
  - killer T-cells release chemicals that punch holes in the cell-surface membrane, so destroying the infected cells.
  - memory T-cells remain in the blood and respond rapidly if the antigen is met again
- During the primary response, to an initial infection, the host suffers disease symptoms. The secondary immune response, to a subsequent infection, is quicker and produces a higher concentration of antibodies so that the pathogen is dealt with before disease symptoms occur.
- Immunity can be stimulated artificially by vaccination. Since this is active in stimulating the production of plasma cells and memory cells, it provides long-term immunity.
- Antibodies can pass naturally into a baby (across the placenta and in mother's milk) or artificially via an injection to counter toxins (e.g. in a snake bite). Since this is passive, it is a short-term response only.
- Transplanting tissue is only possible if the tissue is appropriately matched and medical techniques are used to suppress the immune system.
- Transfusion of blood is only possible if the recipient does not have antibodies complementary to the antigens on the red blood cells of the donor.

Coordination and control

Coordination in flowering plants

Plants are able to detect environmental stimuli. These include:

- the direction of light
- the duration of light (photoperiod)

Plants respond to these environmental stimuli in terms of their growth and development. These coordinated responses allow plants to receive the maximum amount of light for photosynthesis and development (e.g. flowering) at a time of the year that is most appropriate.

Plant growth substances

Most plant responses are controlled by hormone-like chemical coordinators. These usually exert their effect by controlling plant growth, so are commonly called **plant growth substances**.

An outline of plant growth: During the growing season, cell division takes place at the tip (apex) of the plant in tissue called the **apical meristem**. These cells enlarge in a region below the tip called the **zone of elongation**; at certain times of the year the **internode** — the region between the nodes, or points at which leaves develop — also elongates (see Figure 13).

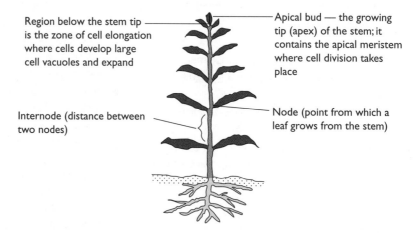

Region below the stem tip is the zone of cell elongation where cells develop large cell vacuoles and expand

Apical bud — the growing tip (apex) of the stem; it contains the apical meristem where cell division takes place

Internode (distance between two nodes)

Node (point from which a leaf grows from the stem)

Figure 13 The main growing regions of a flowering plant

The main effects of the major plant growth substances are shown in Table 6.

Table 6 The main effects of three plant growth substances

Plant growth substances	Site of production	Main plant growth effect
Auxins	Growing tip of stem (apical meristem)	Stimulate elongation of cells (in the zone of elongation)
Cytokinins	Actively dividing (meristematic) tissues	Promote cell division
Gibberellins	Apical buds and leaves	Stimulate elongation of internodal regions

Knowledge check 18

Suggest why only certain tissues in a plant will respond to a particular plant growth substance.

An outline of cell enlargement: The cells that divide by mitosis in the apical meristem are small, have thin cell walls and prominent nuclei, and lack large vacuoles. In the zone of elongation below, these newly formed cells develop small vacuoles and absorb water by osmosis, and eventually large, permanent vacuoles are formed. As the vacuoles absorb water, the cells increase in size — they enlarge (see Figure 14).

Figure 14 The process of cell enlargement

An outline of auxin action:

- Auxins are produced in the cells of the apical meristem.
- They diffuse down the shoot towards the zone of elongation.
- They bind to specific receptors on the cell surface membranes of the newly formed cells.
- This causes the membrane pumps to move hydrogen ions out into the cellulose cell wall.
- This acidification of the cell wall activates agents which loosen the linkages between cellulose microfibrils, allowing slippage between them, and making the wall more flexible.
- The cells absorb water by osmosis and the flexible cell walls allow the cells to expand as the extra water exerts increased hydrostatic pressure against them.
- The more auxin that is received in the zone of elongation, the more this effect allows cells to expand.

Auxins and phototropism

Shoots grow towards light, i.e. they are **positively phototropic**. Basically, light from one direction causes the lateral movement of auxin from the illuminated side of the shoot to the shaded side. Since elongation of cells is stimulated by relatively high concentrations of auxin, the cells on the shaded side grow more. A differential growth response results in positive phototropism. The mechanism is summarised in Figure 15.

Figure 15 A summary of the mechanism of positive phototropism in a shoot

Examiner tip

Don't confuse *tropic* with *trophic*. Tropic describes a growth response towards or away from a stimulus (e.g. phototropic response). Trophic is connected with how living things feed (e.g. plants are autotrophic).

Knowledge check 19

When illuminated from above, the amounts of auxin on the left and right sides of a shoot were measured as 23.8 and 24.3. When illuminated from the right, the amounts of auxin on the left and right sides of a shoot were measured as 35.5 and 12.6. What can you conclude from these results?

Examiner tip

Develop your understanding of phototropism by first knowing a few facts about auxins: they are produced in the tip and affect a response in the zone of elongation, move away from the source of light, and stimulate cell elongation. Knowing these few facts, you should be able to describe the process of phototropism and interpret many of the experiments on tropic responses.

Several noted biologists have undertaken experiments on auxins and phototropism. These experiments used coleoptiles (the protective sheath found on some seedlings). Some of these experiments are explained in Table 7.

Table 7 Experiments used by some noted biologists to investigate phototropism in coleoptiles

Knowledge check 20

Explain how the results in Table 7 support Darwin's conclusion that light is detected by the tip of a coleoptile.

Knowledge check 21

Explain how the results in Table 7 support Boysen-Jensen's conclusion.

Examiner tip

Examiners will not expect you to recall any of these experimental results — they are not detailed in the specification. Examiners would, however, expect you to be able to interpret and evaluate these results — testing AO3.

Biologist (and year of experiment)	Experimental procedure	Observed result	Suggested explanation
Darwin (1880)	Unilateral light; Intact oat coleoptile	Coleoptile bends towards light	The coleoptile is positively phototropic. It bends towards the light by unequal elongation of the region just behind the tip.
	Unilateral light; Coleoptile tip removed and discarded	No response	The tip must either perceive the stimulus or produce the messenger (or both), as the tip removal prevents any response.
	Unilateral light; Lightproof cover is placed over intact tip of coleoptile	No response	The tip of the coleoptile perceives light.
Boysen-Jensen (1913)	Unilateral light; Mica inserted on shaded side	No response	Auxin moves to the shaded side of the coleoptile but is prevented by the mica from moving further down. No auxin moves into the zone of elongation on either side, so no bending occurs.
	Unilateral light; Gelatin block; Tip removed, gelatin block inserted and tip replaced	Coleoptile bends towards light	As gelatin allows chemicals to pass through it, the bending which occurs must be due to a chemical, e.g. auxin, passing from the tip.
Went (1928)	Darkness; tip cut off, placed on agar block for some time and then block placed to side of decapitated coleoptile	Coleoptile bends away from the side on which the agar block is placed	The auxin from the tip is collected in the agar block. When the block is placed to one side on a decapitated coleoptile, the auxin moves down that side, increasing growth and causing bending. The degree of curvature is proportional to the amount of auxin.

The significance of positive phototropism is that it ensures that plants receive as much light as possible for photosynthesis.

Phytochromes and the control of flowering

Some plants flower whenever they have grown sufficiently, irrespective of daylength (photoperiod). However, many flowering plants are sensitive to daylength and are classed according to the photoperiod in which they flower:

- **long-day plants** (LDPs), e.g. poppy (*Papaver* sp.), flower only if the daylength exceeds a critical value.
- **short-day plants** (SDPs), e.g. chrysanthemum, flower only if the daylength is less than a critical value

The photoperiod is measured by **phytochrome pigments**, found in minute quantities in leaves. Phytochrome occurs in two interchangeable forms: **P_{660}** and **P_{730}**. P_{660} (P_R) maximally absorbs red light of wavelength 660 nm, while P_{730} (P_{FR}) maximally absorbs far-red light of wavelength 730 nm. Absorption of light by one form of phytochrome causes its rapid conversion to the other type. During darkness there is a slow conversion of P_{730} to P_{660}, which means P_{660} accumulates at night. Since daylight contains more red light than far-red light, P_{730} accumulates during the day. This information is summarised in Figure 16.

Figure 16 The interconversions of phytochromes (a) experimentally, and (b) in nature

P_{730} *is the physiological active form*, and can operate to either stimulate or inhibit flowering. The amount of P_{730} in the leaves is critical to flowering, and what influences that amount is the *critical length of continuous (uninterrupted) darkness*. The flowering requirements of LDPs and SDPs are summarised in Table 8.

Table 8 A summary of the flowering requirements of long-day and short-day plants

Class of plant	Required photoperiod for flowering	Relative level of P_{730}	Effect of P_{730}
Long-day plant (LDP)	Critically short period of darkness, i.e. shorter than a certain value	High level of P_{730}	Stimulatory
Short day-plant (SDP)	Critically long period of darkness, i.e. longer than a certain value	Low level of P_{730}	Inhibitory; removal of P_{730} required for flowering

The effect of different photoperiods, or light regimes, on the levels of the two forms of phytochrome and the subsequent effect on the flowering of long-day and short-day plants is illustrated in Table 9.

Examiner tip

Know that P_{730} converts to P_{660} (slowly) in darkness and so is removed during a long night (short day); and know that P_{730} is the active form, either stimulating or inhibiting. You can then reason that an LDP flowers when the night is short (long day) so that sufficient P_{730} remains to promote flowering; while an SDP flowers when the night is long (short day) which is when P_{730} is removed, removing an inhibitory effect on flowering.

Knowledge check 22

A long-day plant normally flowers when the dark period becomes critically short. Explain what would happen if, before flowering was induced, the light period (daylength) was interrupted by a short period of far-red light.

Table 9 The effect of different photoperiods on phytochromes and flowering in LDPs and SDPs

Photoperiod 0 ⟵ hours ⟶ 24	Phytochrome response	Effect in LDP	Effect in SDP
Long day, Short night	P_{730} accumulates during the long day, and is not sufficiently removed during the short night — mostly P_{730}	Flowering	No flowering
Short day, Long night	P_{730} is removed during the long night and is not produced sufficiently during the short day — little P_{730}	No flowering	Flowering
Short day, Long night interrupted by short light period	During the night, the short light period converts P_{660} to P_{730} — enabling sufficient P_{730} to accumulate	Flowering	No flowering

Manipulation of the photoperiod enables commercial growers to bring plants to the consumer at the time when they will command high prices, such as at Christmas. Thus chrysanthemums — short-day plants which usually flower in the autumn, as the days shorten — can be prevented from flowering by maintaining them in a long-day lighting regime. Understanding the phytochrome system allows further manipulation. As Table 9 illustrates, exposing the plants to a short period of light during the night can inhibit flowering. In this way flowering can be delayed for several months, until it is induced by a return to uninterrupted long nights — as when chrysanthemums can be readied for the market in time for Christmas.

The phytochrome system operates in leaves. Flowers develop in the terminal or axillary buds. These may be some distance away. This suggests the involvement of a chemical coordinator; however, as yet no such chemical has been isolated.

Synoptic links

You should have an understanding of the general structure of the plant cell, the structure of cellulose and the process of mitosis, which you will find in the AS Unit 1 Student Unit Guide.

Comparison of coordination in plants and animals

Coordination in both plants and animals involves receptors, a linking (communication) system and effectors.

- **Receptors** receive the **stimulus**: for example, in plants the phytochrome system in the leaf detects the photoperiod; in animals, the osmoreceptors in the hypothalamus are stimulated by changes in the water potential of the blood, while the retina of the eye is stimulated by light.

- **Effectors** bring about the **response**: for example, in plants a bud may develop into a flower, while cells elongate in the zone of elongation in response to auxin; in animals, muscle contracts — in the gut to bring about peristalsis, or in a limb to move one bone with respect to another.

Both *plants and animals use chemicals to coordinate*. For example, plants produce auxin in the tip and this moves down through the shoot. In animals, **hormones** are produced in one part of the organism and travel to other parts where they have their effects; for example, ADH is produced by neurosecretory cells in the hypothalamus (and stored in the posterior pituitary) and carried in the blood to the kidney, where it causes the collecting ducts to be permeable to water.

It takes time for chemicals to be produced and take effect, so there is a delay before the response takes place. The advantage is that the response can be maintained over a period of time.

However, animals need an additional system which is fast in linking receptors (e.g. in the retina of the eye) with effectors (e.g. muscles in the leg). This is because animals differ from plants in that they can move from one place to another (**locomotion**). This ability probably developed because animals have to search for food — unlike plants, which are autotrophic. Animals have a **nervous system** containing neurones, which transmit impulses very rapidly.

Summary

- Plant growth regulators include auxins, cytokinins and gibberellins; auxins stimulate the elongation of cells; cytokinins promote cell division; gibberellins stimulate elongation of the internodal regions.

- Auxins are produced in the shoot tip and diffuse down to the zone of elongation. They stimulate the pumping of hydrogen ions out of the cells and this increases the elasticity of the cell walls so that the cells elongate more during the process of enlargement (vacuolation).

- Auxins are involved in producing the positive phototropic response in plant shoots. As the auxins diffuse down from the tip, unidirectional light causes auxin to move towards the shaded side promoting greater cell elongation on the shaded side and a resultant curvature of the shoot towards the light.

- Plants have phytochrome pigments that detect the photoperiod. They are interchangeable: P_{660} is converted in the light (red light) to P_{730}, which slowly converts in the dark (rapidly by far-red light) back to P_{730} (the active form). The phytochrome system is involved in the control of flowering: in long-day plants an accumulation of P_{730} stimulates flowering; in short-day plants P_{730} inhibits flowering and so its removal (during a critically long night) results in flowering.

- In both plants and animals, chemicals are used to communicate between receptors (receiving stimuli) and effectors (effecting responses). However, plants lack a nervous system, a feature of animals that allows rapid communication and is important in coordinating movement.

Coordination in mammals

The mammalian nervous system can be divided into a **central nervous system** (CNS) and a **peripheral nervous system** (PNS). The CNS consists of the **brain** and **spinal cord**. It is responsible for integrating the activity of the nervous system in coordinating the functioning of all parts of the body. The peripheral nervous system consists of **cranial nerves** that are attached to the brain and **spinal nerves** that are attached to the spinal cord. These nerves connect receptors to the CNS and the CNS to effectors.

The **neurone** (nerve cell) is the functional unit of the nervous system. The human brain comprises tens of millions of neurones, each linked to other neurones. A **nerve** is a bundle of neurones.

Neurones and impulse propagation

The structure of neurones

Neurones link different cells and are able to conduct impulses between them. While variable in size and shape, all neurones have three parts:

- a cell body (**centron**), which contains the nucleus and other organelles, and has a number of cytoplasmic extensions
- extensions (called **dendrons** or, if they are very small and numerous, **dendrites**), which transmit impulses to the cell body
- extensions (**axons**), which transmit impulses away from the cell body and terminate in **synaptic bulbs** (or knobs) — axons may be over a metre long

Knowledge check 23

Where would the cell body of a motor neurone be located?

Two types of neurones of the PNS are shown in Figure 17: a sensory neurone, which conducts an impulse from a receptor to the CNS; and a motor neurone, which conducts an impulse from the CNS to an effector.

Knowledge check 24

What is the name of the supporting cell that produces the myelin sheath?

Figure 17 The structure of (a) a sensory neurone, and (b) a motor neurone

CCEA A2 Biology

In mammals, many axons and dendrons are **myelinated** — covered with a **myelin sheath**. The myelin sheath is made up of many layers of cell membrane of **Schwann cells**, which wrap themselves round and round the axon. This insulates the axon, preventing ion movement between the axon and the tissue fluid around it. Each Schwann cell forms a sheath about 1mm long. There are small spaces between Schwann cells, giving rise to gaps in the myelin sheath called **nodes of Ranvier**. At these nodes the axon is exposed.

The resting potential

All cells have a potential difference across their cell surface membranes, due to differences in the distribution of ions. Neurones are special in having a particularly large potential difference. This is essentially due to a large excess of positively charged sodium ions (Na^+) on the outside. Since this potential difference occurs when the neurone is 'at rest', it is called the **resting potential**. The inside of the neurone is negative with respect to the outside and the magnitude of the resting potential is approximately $-70\,mV$ (millivolts).

The action potential

Neurones (and also muscle and receptor cells) are excitable — the potential difference can be reversed. When a neurone is stimulated, the cell-surface membrane allows sodium ions to diffuse in. This changes the potential difference across the membrane, making it less negative inside. If a critical potential difference (of about $-55\,mV$), called the **threshold value**, is reached, then ions surge in and the neurone quickly becomes **depolarised**. Indeed, the cell will become positive on the inside, reaching a potential difference of about $+40\,mV$. This reversal of the potential difference is called an **action potential**. An action potential does not vary in size, and either occurs (if the threshold value is achieved) or does not, a phenomenon called the **all-or-nothing law**. The action potential is followed by a period when the membrane repolarises and recovers its resting potential. This recovery period is called the **refractory period**; during this time the membrane is unexcitable. These changes are shown in Figure 18.

① Axon membrane at resting potential
② Depolarisation to threshold value needed for further depolarisation to occur
③ Further depolarisation leads to action potential
④ Magnitude of action potential: 40 mV − (−70mV) = 110mV
⑤ Repolarisation of axon membrane during the refractory period when the membrane cannot be depolarised

Figure 18 Changes in the potential difference across the axon membrane as an action potential occurs and as it recovers its resting potential (note that these are changes over time at one point on the membrane)

Examiner tip

A membrane is said to be *polarised* if a potential difference is maintained across it with the inside being negative with respect to the outside. *Depolarisation* is a decrease and then temporary reversal of the potential difference so that the inside of the membrane becomes positive with respect to the outside. The membrane is actually still polarised but the *polarity is reversed*.

Examiner tip

While the specification states that the ionic basis for potential differences is *not* required, you would benefit from understanding that: during the resting potential there is an excess of Na^+ outside a membrane that is impermeable to it; during the action potential the membrane becomes permeable and Na^+ diffuses in; and that during repolarisation Na^+ is pumped out. It is the detail with respect to other ions and the functioning of the channel and carrier proteins that is not required.

Examiner tip

Local circuits are the movements of ions along the neurone. The flow of ions is caused by an increase in concentration at one point, which causes diffusion away from the region of higher concentration.

Knowledge check 26

Suggest why a refractory period is important for neurone function.

Examiner tip

The nerve impulse is a wave of depolarisation. This is similar to the crowd in a sports stadium performing a 'Mexican wave'. At any one time only one section of the crowd is waving. The people do not move along, they stay in one place, but the wave travels along the crowd from one section to another. Similarly, one action potential does not actually travel along the neurone; a series of action potentials form *in turn* along the neurone.

Knowledge check 27

Suggest a disadvantage of the nodes of Ranvier being (a) closer together and (b) further apart.

Impulse propagation

An action potential that is generated in one part of a neurone is propagated rapidly along its dendron or axon. This happens because the depolarisation of one part of the membrane sets up **local circuits** with the areas either side of it. Local circuits occur as positive ions are attracted by neighbouring negative regions and flow in both directions. On one side, the membrane is still recovering its resting potential (repolarising), i.e. it is in its refractory period during which it cannot be stimulated. On the other excitable side, the local circuit triggers depolarisation and the formation of an action potential (see Figure 19). This process is repeated at each section of the membrane along the whole length of the neurone. In essence, then, an *impulse is the transmission or propagation of depolarisations*, or action potentials, along the neurone membrane.

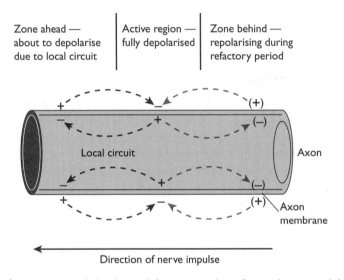

Figure 19 Local circuits and the propagation of an action potential

In a myelinated neurone (such as a sensory or motor neurone in mammals), local circuits cannot be set up in the parts of the neurone insulated by the myelin sheath. Instead, the *action potential 'jumps' from one node of Ranvier to the next*. This greatly increases the speed at which it is propagated along the axon and is called **saltatory conduction**. Some non-myelinated axons have a large diameter, and so a larger surface over which ions might be moved; thus the formation of action potentials occurs more rapidly and impulse transmission is speeded up.

The synapse and synaptic transmission

At the end of an axon, the impulse will arrive at the synaptic bulbs. These come very close to their target cells (another neurone or muscle cell). There is a small gap between them called a synaptic cleft. The membrane of the neurone just before the cleft is called the pre-synaptic membrane, and the one on the other side is the post-synaptic membrane. The whole structure is called a synapse (see Figure 20).

The synaptic bulbs possess synaptic vesicles which carry the neurotransmitter chemical; in the peripheral nervous system, **acetylcholine** is the main

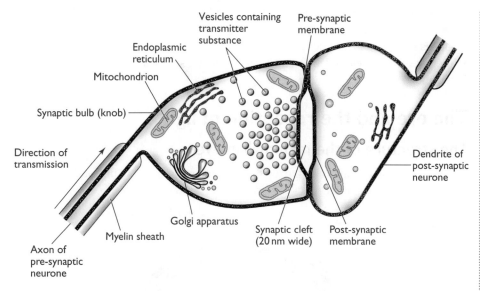

Figure 20 The structure of a synapse

neuro-transmitter, though some synaptic bulbs contain **noradrenaline**. The sequence of synaptic transmission is as follows:

- When an impulse arrives at the synaptic bulb, the membrane becomes permeable to calcium ions which diffuse into the bulb.
- Calcium ions stimulate movement of synaptic vesicles towards the pre-synaptic membrane.
- Vesicles fuse with the pre-synaptic membrane and release acetylcholine (ACh) molecules by exocytosis into the synaptic cleft.
- Acetylcholine molecules diffuse across the synaptic cleft to the post-synaptic membrane; because the gap is only 20 nanometres (nm) wide, this takes only a fraction of a millisecond (ms) to occur.
- On the post-synaptic membrane, acetylcholine molecules attach to receptors.
- This causes the post-synaptic membrane gradually to become depolarised (as sodium ions diffuse in) so that an excitatory post-synaptic potential (EPSP) is generated.
- The degree of depolarisation (and the EPSP) is dependent on the amount of transmitter released and so the number of receptors filled.
- If the EPSP reaches a threshold intensity then an action potential is evoked, and an impulse is propagated in the post-synaptic cell.
- Acetylcholine molecules are hydrolysed by the enzyme acetylcholinesterase and the breakdown products, choline and ethanoic acid, released into the cleft — this breakdown of Ach is important as it prevents the continued stimulation and firing of impulses in the post-synaptic neurone.
- The breakdown products diffuse across the cleft and are reabsorbed into the synaptic bulb where they are resynthesised into acetylcholine using energy in the form of ATP from mitochondria.

Even though synapses *slow transmission* slightly, there are distinct *advantages*. They:
- ensure that transmission is in one direction only, since synaptic knobs only occur at one end of a neurone and the neurotransmitter is only released from the pre-synaptic side

Knowledge check 28

Explain the role of the (a) the Golgi apparatus and (b) mitochondria in the synaptic bulb.

Knowledge check 29

Which mineral ion causes synaptic vesicles to fuse with the pre-synaptic membrane?

Knowledge check 30

A number of drugs or poisons exert their effect at synapses. Suggest the effect of: (a) the South American arrow poison curare, which blocks the acetylcholine receptors on the post-synaptic membrane; (b) the nerve gas sarin, which inhibits the enzyme acetylcholinesterase.

- protect effectors (muscles and glands) from over-stimulation, since continuous transmission of action potentials exhausts the supply of transmitter substance, i.e. causes synapses to fatigue
- integrate the activity of different neurones synapsing with a single postsynaptic neurone

The eye and the reception of light

The structure of the mammalian eye

The structure of the mammalian eye is shown is Figure 21. The main parts are described, along with their functions, in Table 10.

Figure 21 The structure of the mammalian eye

Table 10 The main parts of the mammalian eye and their functions

Part of eye	Description	Function
Conjunctiva	Very thin transparent membrane covering the cornea and lining the eyelids	Protects the cornea at the front of the eyeball
Sclera	Tough connective tissue that encloses the eye, except at the front	Protects the eyeball against mechanical damage; allows attachment of eye muscles
Cornea	The front part of the sclera — transparent to light	Allows the passage of light while refracting (bending) it
Iris	The coloured part of the eye — it contains circular and radial muscles	Controls the size of the pupil to adjust the amount of light entering the eye

Part of eye	Description	Function
Pupil	A gap within the iris — appears black (light goes in and does not come out)	Allows light to enter the eye — its size can be varied
Aqueous humour	Transparent watery fluid filling the front part of the eye	Maintains the shape of the front chamber of the eye
Lens	Transparent and elastic ovoid (biconvex) structure held in place behind the cornea	Changes shape to adjust the focusing (accommodation) of light onto the retina
Ciliary body	Structure which supports the lens and contains circular muscle	Contraction or relaxation of the circular muscle controls the shape of the lens
Suspensory ligaments	Strong ligaments which connect the ciliary body to the lens	Transfers tension in the wall of eyeball to the lens to make the lens thinner
Vitreous humour	Transparent, jelly-like material filling the rear part of the eyeball	Maintains the shape of the rear part of the eyeball and supports the lens
Retina	Inner layer of the wall of the eyeball containing the light-sensitive cells	The light-sensitive cells (rods and cones) initiate impulses in associated neurones when appropriately stimulated
Fovea	The region at the back of the retina that is rich in cones	A region with high visual acuity that allows colour vision
Choroid	A layer of pigmented cells at the back of the eyeball behind the retina	Contains blood vessels which supply the retina; prevents reflection of light within the eyeball
Optic nerve	A bundle of sensory nerve fibres which leave from the back of the eye	Transmits impulses from the retina to the optic centre at the back of the brain
Blind spot	Region where optic nerve leaves inside of eyeball and so contains no light-sensitive cells	A region which, if light strikes it, is not sensitive to light

The iris and the control of pupil diameter

In *bright light*, the **circular muscles** *contract* and the radial muscles relax, making the iris widen and therefore making the pupil get narrower. The *pupil is small* (constricted) to limit the amount of light passing through, as very bright light can damage rods and cones.

In *dim light*, the opposite takes place: circular muscles relax and **radial muscles** *contract*, widening (dilating) the pupil. The *pupil is large* to allow more light to reach the retina and so provide maximal stimulation of the light-sensitive cells there.

The lens and the accommodation (focusing) of light

The cornea refracts light, but it is the lens that adjusts the refraction of light to a single point on the retina. The ability to adjust focusing is called **accommodation**. This occurs by means of adjustments to the shape of the lens.

The wall of the eyeball is under pressure, simply by being filled with fluid. If the **ciliary muscles** *are relaxed*, this pressure is transferred via the suspensory ligaments to the lens, pulling it into a *thin shape*. This means that the lens does not converge the light as much, which is what is required to accommodate light onto the retina from a *far object*.

Knowledge check 31

What is the function of the cornea?

Examiner tip

Don't try to just memorise how the iris functions; make sense of it. In *bright light* a *small pupil* is required (to prevent photoreceptor damage) so the iris is pulled towards the centre by *contraction of circular* muscle; in *dim light* a *big pupil* is required (to sufficiently stimulate photoreceptors) so the iris is pulled away from the centre by *contraction of radial* muscle.

Examiner tip

For focusing onto a far object, the lens is pulled into a thin shape. Don't say that this is due to *contraction* of the suspensory ligaments. The suspensory ligaments simply transfer the tension in the wall of the eye through to the lens.

To accommodate light onto the retina from a *near object*, the *circular muscles in the ciliary body contract*. This closes the aperture around the lens and releases any tension from the eyeball. The lens, being elastic, adopts a *fatter shape*, which means that it refracts the light more, so accommodating light onto the retina.

The structure of the retina

The retina of the human eye contains two types of cells which are receptors to light: **rods**, which are sensitive to dim light; and **cones**, which only respond in bright light but which are able to discriminate fine detail and distinguish different wavelengths (colours) of light. Cones are concentrated at the fovea, at the centre of the retina (see Figure 21), while rods are mainly found around the periphery of the retina. The rods and cones synapse with bipolar neurones which, in turn, synapse with neurones of the optic nerve. Significantly, many rods synapse with each bipolar neurone and many bipolar cells connect with each neurone of the optic nerve. This is called **retinal convergence**. Cones, on the other hand, generally synapse with a single bipolar neurone and a single neurone of the optic nerve. Note that the light passes through the neurones before reaching the outer segments of the rods and cones. The consequence of this is that for the neurones to leave the eye they must pass through the layer of photoreceptors, creating an area devoid of receptors — the blind spot. The retina is shown in Figure 22.

Figure 22 The structure of the retina

CCEA A2 Biology

Rods and vision in dim light

The outer segment of a rod cell contains many membranes, all stacked up in parallel to one another. These membranes contain a pigment called **rhodopsin**, which is composed of a protein (opsin) and a light-absorbing compound derived from vitamin A, called retinal. When light strikes rhodopsin, changes occur, and rhodopsin breaks down into retinal and opsin. This results in a change in the membrane potential of the rod cell and creates a **generator potential**. This causes a change in the membrane potential of the neighbouring bipolar neurone, with which the rod cell synapses. The bipolar cell releases transmitter substance into its synapse with a neurone of the optic nerve. If sufficient transmitter substance is released, an action potential is generated in the neurone of the optic nerve and transmitted to the visual centre at the back of the brain. Rod cells are sensitive, partly because rhodopsin absorbs light readily and is easily broken down and, to a greater extent, because retinal convergence enables the input of many rods to be added together. This **summation** occurs because generator potentials have an additive effect in stimulating bipolar neurones while transmitter substance from bipolar neurones is added to reach the threshold needed to initiate an action potential in the neurone of the optic nerve. However, the consequence of retinal convergence is that the brain cannot distinguish which rod of a group sharing the same optic nerve fibre has been stimulated, i.e. there is decreased visual acuity and rods lack the ability to discriminate detail.

Rhodopsin is continuously re-synthesised within the rods. This re-synthesis requires ATP from the mitochondria. The rate of re-synthesis is sufficient for the rods to continue functioning in dim light. However, in bright light the rhodopsin is almost entirely bleached. It takes about 30 minutes in complete darkness for all the rhodopsin to be re-formed and so for the rods to become functional, a phenomenon called **dark adaptation**.

Cones, visual acuity and colour vision in bright light

Light reception in cones is very similar to that in rods. The pigment involved here is **iodopsin**. Iodopsin is less readily broken down and cones only produce a generator potential in bright light. Furthermore, cones do not exhibit convergence and one cone associates with one neurone of the optic nerve. Each cone cell stimulated will generate an impulse to the brain. This gives high **visual acuity** — the brain is able to distinguish between points that are close together.

Further, iodopsin exists in three different forms and there are three different types of cone, each with a different type of iodopsin. Each is sensitive to different wavelengths of light corresponding to the colours blue, green and red. This forms the basis of the **trichromatic theory of colour vision**. Pure red light will only break down the red iodopsin and only the red cones will fire impulses to the brain. This is interpreted by the brain as red. However, yellow light will break down some of the red iodopsin and some of the green iodopsin and so both red and green cones will fire impulses to the brain. This is interpreted by the brain as yellow. Thus colour perception in the brain depends on the relative proportions of the different types of cone that are firing impulses.

Examiner tip

If asked to explain summation, don't say that impulses are added together — this is simply wrong. What you should say is that, with low-intensity stimuli, generator potentials have an additive effect and that there is enough transmitter released to reach the threshold for an action potential to be generated (and an impulse propagated) in a neurone of the optic nerve.

Knowledge check 33

Explain why objects that appear brightly coloured in a well-lit room appear in dim light to be various shades of grey.

Knowledge check 34

According to the trichromatic theory of colour vision, to which colours of light are the three different types of cone sensitive?

Binocular vision and visual fields

Some two-eyed animals, usually prey animals (e.g. rabbits), have their eyes positioned on opposite sides of their heads so as to give the *widest possible field of view*. A wide visual field facilitates the detection of potential predators. Other two-eyed animals, usually predatory animals or primates (including humans), have eyes positioned on the front of their heads. The use of both eyes to view an object (**binocular vision**), creating a single image, allows more accurate *judgement of distance*. It also allows **stereoscopic vision** as a result of an increase in *depth perception*, so that the brain can create a three-dimensional image.

Knowledge check 35

Distinguish between binocular and stereoscopic vision.

Practical work

Examine prepared slides or photographs of the mammalian eye:

- Recognise the following components — conjunctiva, cornea, iris, pupil, ciliary body, suspensory ligaments, aqueous and vitreous humours, retina, choroid, sclera, blind spot, optic nerve, rods and cones

Muscle and muscle contraction

Muscle is an 'excitable' tissue and is capable of contraction. There are three main types of muscle: **skeletal**, **cardiac** and **smooth**. These are compared in Table 11.

Table 11 A comparison of cardiac, smooth and skeletal muscle

	Skeletal	Cardiac	Smooth
Appearance	Muscle fibres are multinucleate, with distinct striations (bands)	Cells are striated and branched, forming a linked network; intercalated discs between cells	Spindle-shaped cells with a single nucleus and no striations
Distribution	Attached by tendons to bones	Only found in the wall of the heart	Present in the iris and ciliary body of the eye, and in walls of tubular organs, e.g. gut, blood vessels and bladder
Function	Movement of parts of the body and locomotion; strong contractions but not long-lasting	Pumping of the heart maintains blood circulation; cardiac muscle will contract throughout life	Movement of materials within the body; peristalsis in the gut can continue for long periods

Knowledge check 36

List the three types of muscle in order according to their tendency to fatigue, starting with the muscle type that fatigues least.

Examiner tip

The different arrangements of myofibrils, in relaxed and contracted muscle fibres, often come up in exams. You need to be able to interpret myofibril (actin and myosin) arrangements in longitudinal and transverse sections, in both relaxed and contracted states.

The structure of skeletal muscle

Skeletal muscle consists of bundles of muscle fibres. A muscle fibre is multinucleate, resulting from the fusion of many cells, so is relatively large, e.g. up to $100\,\mu m$ in width and as long as $300\,mm$. The nuclei lie just beneath the **sarcolemma** (the surface membrane), out of the way of the packed **myofibrils**, each of which is surrounded in **sarcoplasmic reticulum** and joined transversely by **T-tubules**, and between which there are numerous mitochondria. Each myofibril consists of overlapping arrays of thick and thin filaments composed, respectively, of the proteins **myosin** and **actin**.

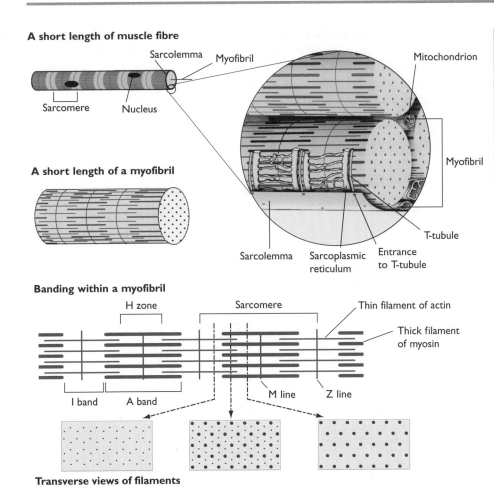

A short length of muscle fibre

Sarcolemma Myofibril Mitochondrion

Sarcomere Nucleus

Myofibril

A short length of a myofibril

T-tubule

Sarcolemma Sarcoplasmic reticulum Entrance to T-tubule

Banding within a myofibril

H zone Sarcomere Thin filament of actin

Thick filament of myosin

M line Z line

I band A band

Transverse views of filaments

Figure 23 The structure of a muscle fibre

Knowledge check 37

How does the appearance of each of the following alter when a muscle becomes contracted: I band, H zone, A band and sarcomere? Does it become longer, shorter or stay the same?

Muscle contraction

The shortening of myofibrils, according to **sliding filament theory**, causes muscle contraction. The sequence of myofibril shortening is as follows:

- An action potential arrives via a motor neurone at the synapse (neuro-muscular junction) with the cell surface membrane (sarcolemma) of the muscle fibre.
- Action potentials are propagated through the T-tubules and along the sarcoplasmic reticulum, causing calcium ions (Ca^{2+}), which are stored in the sarcoplasmic reticulum, to be released into the cytoplasm (sarcoplasm).
- Calcium ions cause ancillary proteins, which normally cover binding sites on the actin filaments, to be displaced and so uncover the binding sites.
- Heads of the myosin molecules next to the uncovered binding sites now attach to the actin filaments, forming acto-myosin 'bridges' between them.
- The myosin heads rotate or 'rock' back, pulling the thin actin filaments over the thick myosin filaments.
- ATP now binds with the myosin heads and the energy released from its hydrolysis (by ATPase) causes the myosin heads to detach from the actin filaments.

- The detached myosin heads regain the original position, and attach to another exposed binding site on the actin filament, so that the cycle of attachment, rotation and detachment is repeated.
- This process continues as long as action potentials are propagated through the muscle fibre.

Figure 24 illustrates how the sliding of actin filaments over myosin filaments causes the myofibril, and the muscle fibre, to shorten.

Figure 24 Movement of the myosin head causes the actin filament to slide over the myosin filament

Practical work

Examine prepared slides or photographs of skeletal muscle, cardiac muscle and smooth muscle:
- Recognise the characteristic features of skeletal muscle and cardiac muscle using LM and TEM photographs and smooth muscle using LM photographs

Synoptic links

You should have an understanding of the general structure of an animal cell, which you will find in the AS Unit 1 Student Unit Guide. This will allow you to appreciate just how specialised nerve, photoreceptor and muscle cells are. You should also have an understanding of cell ultrastructure and of the structure of cell membranes.

Neurones and synapses:

- Neurones (sensory and motor) have a centron and long processes, dendrons or axons, which may have associated Schwann cells (forming a myelin sheath) separated by nodes of Ranvier.

- A non-conducting neurone has a resting potential across its membrane whereby it is negative on the inside.

- When stimulated, the neurone membrane is depolarised and becomes positive on the inside as an action potential develops.

- A nerve impulse consists of a wave of action potentials (depolarisations) moving along an axon.

- The nerve impulse travels faster in myelinated neurones as it jumps between the nodes of Ranvier (saltatory conduction). Impulses also travel faster along axons with larger diameters.

- When an impulse arrives at the end of a neurone, neurotransmitter is secreted into the synaptic cleft, diffuses across and attaches to protein receptors on the post-synaptic membrane, which is depolarised as an excitatory post-synaptic potential. If this reaches threshold, an action potential is generated and an impulse propagated in the post-synaptic neurone.

The eye:

- The eye possesses the photoreceptor cells, rods and cones, within the retina.

- An iris controls the cone of light entering the eye: circular muscles contract to constrict the pupil; radial muscles contract to dilate the pupil.

- The lens accommodates light onto the retina: tension in the wall of the eyeball pulls the lens into a thin shape (for far objects); contraction of ciliary muscle relieves this tension and the lens becomes fatter (for near objects)

- Rods and cones contain photosensitive pigment, which is broken down by light to produce a change in potential difference. If this is large enough it generates an action potential and an impulse passes via the optic nerve to the brain.

- Rods and cones connect with the neurones of the optic nerve differently. Rods show convergence whereas cones do not. As a result, rods are more sensitive to dim light; cones have less sensitivity but produce an image with greater detail.

- In the trichromatic theory of colour vision, there are three types of cone (red, green and blue) sensitive to different wavelengths.

Muscle contraction:

- Mammals have three types of muscle: smooth, cardiac and skeletal (striated).

- Skeletal muscle consists of muscle fibres, which are multinucleate and possess many mitochondria, extensive sarcoplasmic reticulum (for storage of calcium ions) and infoldings of the cell-surface membrane called T-tubules.

- Muscle fibres comprise myofibrils consisting of actin (thin) and myosin (thick) filaments. Each myofibril has recognisable repeating units called sarcomeres, each of which contains overlapping filaments of actin and myosin.

- Contraction of the sarcomere (and so muscle) involves the following: the release of calcium ions unblocks binding sites on the actin filaments; myosin heads attach to the actin and the acto-myosin bridges rotate, causing filaments to slide over each other; ATP allows the myosin to detach and so the cycle of attachment–rotation–detachment is repeated.

Ecosystems

Populations

A **population** is a group of organisms of the same species living in a particular habitat. As members of a population are of the same species, they will:

- reproduce as long as there are available **resources** (any substance consumed by an organism, e.g. nutrients)
- compete (**intra-specific competition**), if the resources are in limited supply

Phases of population growth

For a new population starting in a particular area (e.g. in a laboratory culture or colonising a new habitat), four phases of growth are recognised. These are shown in Figure 25.

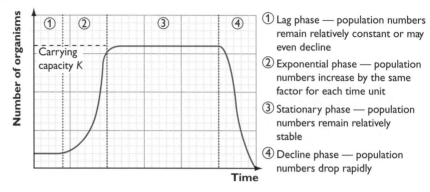

Figure 25 The phases of population growth

Outline of the events in each phase:

- **Lag phase** — this is the time for nutrient assimilation, and/or egg production, and/or egg and larval development, or gestation period (in mammals).
- **Exponential (log) phase** — reproduction creates new members of the population which add to the population's reproductive capacity (e.g. a bacterium divides to produce two bacteria, each of which produces two to give a total of four bacteria, and so on), and there is little competition since there are sufficient resources. The numbers increase by a value called the **intrinsic rate of natural increase** (designated r) and the population is illustrating its **biotic potential** (the reproductive capacity of a population under optimum environmental conditions).
- **Stationary phase** — as population numbers increase so resources become limiting and competition increases (and/or there is accumulation of waste, particularly in laboratory populations), i.e. there is **environmental resistance** (which prevents the population reaching its biotic potential); there is a decline in birth rate combined with an increase in death rate so that recruits (via births, and possibly immigration) equate with losses (via deaths, and possibly emigration), and the population size is determined by the resources available in the environment; the population reaches its **carrying capacity** (designated K), which is the maximum number that the environment can support.
- **Decline phase** — the population has exhausted the resources and/or there is an accumulation of toxic waste so that the birth rate falls to zero and the death rate increases; population numbers crash.

Renewable and non-renewable resources: The stationary and decline phases will depend on whether the resources are renewable (continually being replaced and made available to the organisms) or non-renewable (initially available but not replaced).

- If resources are **renewable** (e.g. in woodland, trees continually fall and provide food for woodlice), the population will tend to remain in its stationary phase.
- If resources are **non-renewable** (e.g. yeast grown in a laboratory batch culture), the population will have an exponential phase followed by rapid decline as the resources are consumed.

Factors influencing populations may be grouped into two main categories:
- **abiotic factors** — factors in the chemical and physical environment, e.g. carbon dioxide concentration, oxygen concentration, availability of mineral ions, water, light, temperature
- **biotic factors** — the effects of other organisms of the same or different species

Intra-specific competition: Competition between members of the same population will become more severe as the population increases and resources become limiting.

The influence of temperature: Temperature is not a resource but will determine the metabolic rate in organisms and so the rate at which they develop. For example, in laboratory populations it can be demonstrated that the rate of increase (in the exponential phase) will rise at higher temperature, but that a higher temperature will not influence the size of the maximum population (in the stationary phase); this will be determined by resources such as available nutrients. Similarly, a warm spring will produce rapid increases in insect populations (beneficial for the growth of insectivorous bird populations).

Population dynamics

The number of individuals making up natural populations fluctuates over time: populations gain individuals through births or immigration (movements into the population), and lose individuals through deaths and emigration (movements out of the population). **Population growth** is determined by the equation:

population growth = births (B) – deaths (D) + immigration (I) – emigration (E)

A population in equilibrium then will have an equation: $B + I = D + E$.

Population dynamics: r-selected and K-selected species

Population dynamics of different species: Two types of species are identified with respect to their reproductive strategy and the dynamics of their populations:
- The populations of the species increase rapidly as a resource becomes available and crash as the resource is used up, with repeated cycles of 'boom-and-bust'; because of the prominence of the 'intrinsic rate of natural increase' (r), such species are called **r-selected species** or are said to have an **r-strategy**.
- The populations of the species remain at the carrying capacity of the environment (K); such species are called **K-selected species** or are said to have a **K-strategy**.

Table 12 compares the features of r- and K-selected species. Most species have strategies between the two extremes.

Table 12 The features of r-selected and K-selected species

Feature	r-selected (r-strategist)	K-selected (K-strategist)
Length of life cycle	Short — quick to mature	Long — takes time before individuals become reproductively mature
Generation time	Short	Long
Numbers of offspring	Many	Few

Table 12 (*continued*)

Feature	r-selected (r-strategist)	K-selected (K-strategist)
Population density	Highly variable; often overshoots K, resulting in 'boom-and-bust' dynamics	Less variable; usually near K
Dispersal (ability to migrate)	High; species migrate readily and are able to re-colonise easily	Low; re-colonisation is uncommon
Competitive ability	Weak competitor	Strong competitor
Body size	Small	Large
Amount of parental care	Little	Considerable
Habitat	Unstable or disturbed	Stable and/or stressful

K-selected species are more prone to extinction as they cannot respond well to environmental disaster.

Practical work

Investigate the growth of a yeast population using a haemocytometer:
● components of a culture medium for a yeast population
● use of a haemocytometer to include counting cells over a grid of determined volume and the calculation of cell density

Estimation of the size of an animal population using a simple mark–release–recapture technique:
● marking techniques
● assumptions made when using a mark–release–recapture technique
● estimation of population size using the Lincoln index (Peterson estimate)

Population interactions

Different populations within a habitat may affect each other's population growth. Table 13 shows three interactions: '+' indicates that there is a positive effect on one population, while '−' indicates that there is a negative effect and that the population would decrease.

Table 13 Three types of population interaction

Type of interaction	Effect on population growth	Comment
Mutualism	+/+	Both species gain; interaction may be necessary for both
Predation	+/−	The predator species gains, the prey species loses
Competition	−/−	Both species lose while interacting; the species most affected is eliminated from its niche

Mutualism

If both species benefit from an interaction, their interaction is called **mutualism**. Nitrogen-fixing bacteria of the genus *Rhizobium* receive protection and nutrients

Examiner tip
You may be asked to calculate estimates of population size: of a yeast population using a haemocytometer; of an animal population using the mark–release–recapture technique; of a plant population sampled using randomly placed quadrats. (These may be found in the CCEA Unit 3 guide.)

Knowledge check 43
Parasitism and predation are both +/− interactions. How do they differ?

within the root-nodule tissue of leguminous plants (such as clover, *Trifolium*) and provide the plant with nitrogen-containing compounds. Lichens are compound organisms consisting of highly modified fungi that harbour green algae among their **hyphae**; the fungi absorb water and nutrients and provide a supporting structure, while the algae carry out photosynthesis.

Predator–prey interactions

Predator–prey interactions change according to the relative numbers of prey and predator. An abundance of prey will mean that more of the predator population can be supported and so the predator population would grow. Large numbers of predators will reduce the prey population; few available prey would then not support the large numbers of predators, which would subsequently fall. The resulting predator–prey interaction will often produce oscillations, especially when the predator species hunts only one or a few prey species. Furthermore, the oscillations produced will show changes in the predator-species population which lag behind changes in the prey population (see Figure 26).

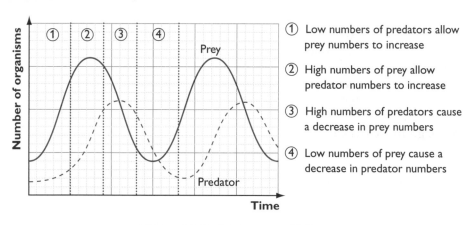

① Low numbers of predators allow prey numbers to increase

② High numbers of prey allow predator numbers to increase

③ High numbers of predators cause a decrease in prey numbers

④ Low numbers of prey cause a decrease in predator numbers

Figure 26 Predator–prey oscillations

Inter-specific competition

Inter-specific competition between the populations of two species occurs when they require a *common resource which is in limited supply*. The species may share the same niche or, at least, exhibit niche overlap. The characteristics of competition are:

- Both species do less well when competing for the resource.
- One species is eventually eliminated from the habitat.
- The winner may utilise the resource more efficiently and so be more successful, or it may have some feature the sole effect of which is to allow the winner to compete more effectively (e.g. the aquatic plant *Lemna gibba* has air sacs which allow it to float above plants of the *L. polyrrhiza* species, so *L. gibba* can absorb more of the available light).
- The outcome of competition may well be determined by the environmental conditions (e.g. the flour beetle *Tribolium castaneum* outcompetes *T. confusum* when conditions are warm and humid but not if it is cold and dry).

Figure 27 shows the population growth curves for two plant species when growing in isolation and when competing.

Knowledge check 44

The roots of many trees may grow in close association with the hyphae of particular fungi: the roots are able to absorb mineral ions released during decay by the fungi; the hyphae are able to absorb sugars from the plant. What type of interaction is being illustrated?

Knowledge check 45

Study Figure 26. Explain why (a) the rise and fall in the predator population always lags behind that in the prey population and (b) the prey population is larger than the predator population (and indeed larger than shown in the graph).

Examiner tip

Don't confuse the terms *intra*-specific and *inter*-specific competition. *Intra-specific* competition occurs between members of the *same* species; *inter*-specific competition occurs between members of *different* species.

Examiner tip

You will need to be able to interpret data in graphs and tables that show changes in population numbers. For example, you should be able to analyse predator–prey and inter-specific competition curves, such as those in Figures 26 and 27. These skills should be practised. Past paper questions are a source of suitable material for practice.

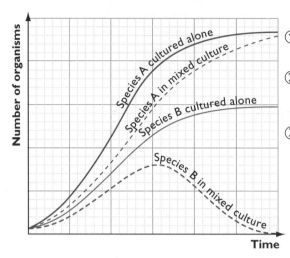

① Both species do less well when competing in mixed culture

② Species A is the winner — it utilises resources better or has an adaptive feature

③ Species B is eliminated

Figure 27 The population growth curves for two competing plant species when grown in pure and in mixed cultures

Knowledge check 46

Why is it not possible for two different species to have identical ecological niches?

The **competitive exclusion principle** is that two species cannot share the same niche without one species being eliminated. However, caution needs to be exercised: two species of grass might require the same mineral ions but obtain these at different levels in the soil because one has longer roots than the other; two species of plant would require light but avoid competition by growing at different times in the year.

Biological control

A **pest** is any organism that competes with or adversely affects a population of plants or animals that are of economic importance to humans. **Biological control** aims to achieve permanent control of pest populations without the dangerous side-effects associated with chemical pesticides. A biological control method involves the introduction of a predator, parasite or pathogen (the **biological control agent**) to reduce the pest numbers to a point below the **economic damage threshold** (see Figure 28).

Figure 28 The principle of biological control

CCEA A2 Biology

Advantages of using biological control agents over the use of pesticides include:

- Biological control agents are more likely to target only the pest species.
- Biological control agents do not have a negative effect on biodiversity.
- The development of resistance of pests to biological control is unlikely.
- Biological control agents reproduce and so are self-perpetuating over time.
- Biological control is relatively cheap since, while the initial research may be costly, the predator or parasite does not have to be reintroduced (whereas pesticides must be re-applied often).

A biological control programme will require much research to ensure a successful outcome; for example, the control agent should target the pest species; it should not itself become a pest; it should survive and reproduce in its new environment; and no diseases should be imported with the control agent.

Biological control cannot cope with sudden surges in pest numbers and pesticides may need to be used in such circumstances.

Knowledge check 47

List the properties of an organism used for the efficient biological control of a pest.

Summary

- A population is all the individuals of a particular species in a particular habitat at a particular time.
- Population growth exhibits four stages:
  - lag phase – the population is low and relatively stable as organisms adapt to their surroundings
  - exponential phase – the population increases rapidly because there are abundant (unlimited) resources
  - stationary phase – the population numbers are high and stable as births are matched by deaths; this is the carrying capacity of the environment
  - decline phase – overuse of resources or accumulation of waste products causes a decline in the population; if resources are renewed, the population may recover, otherwise it crashes to zero
- The factors that limit population growth are known as environmental resistance; such factors can be abiotic or biotic.
- Two types of population growth strategies are recognised:
  - r-selected species show rapid growth, exhausting resources quickly, and move on; a 'boom-and-bust' strategy
  - K-selected species are more conservative; populations grow slowly though numbers remain relatively stable at the carrying capacity, unless disturbed
- Population numbers are influenced by a number of biotic factors:
  - mutualism, where populations of different species benefit from the association
  - predator–prey, where one species (the predator) preys on another; oscillating population curves may result, with predator numbers lagging behind those of the prey
  - inter-specific competition, where two species compete for a common resource; the outcome is the removal of one species by the other
- Biological control is the use of predators, parasites or pathogens to limit the numbers of a pest species; the control species should be highly specific (to the pest), long-term (remain in the ecosystem) and should not contaminate the environment (not itself become a pest).

Communities

A **community** consists of all the populations of organisms in a particular habitat. An **ecosystem** consists of a habitat and its associated community, so that it includes both biotic and abiotic components, and the interactions within and between them.

Community development

Ecosystems constantly change, with new species entering the community and others being lost. **Succession** is the term used to describe the progressive change in the species composition of a community over a period of time. There are two kinds of succession: primary and secondary.

Primary succession

A **primary succession** takes place in a previously uncolonised substrate. Examples include: volcanoes erupting and depositing lava; a glacier retreating and depositing rock; landslides exposing rocks; sand being piled into dunes by sea or wind; lakes being created by subsiding land.

A small number of **pioneer** plants, which are usually r-strategists specialised in dispersal to and colonisation of exposed areas, dominate new communities. This simple community modifies the abiotic environment (causes it to change). This results in a change in the community, which further alters the abiotic environment, and so on. Each successive community makes the environment more favourable for the establishment of new species. The process continues through a number of stages (called **seres**) until a final stage, the **climax community**, is formed. The climax community is a relatively stable end-stage and is in dynamic equilibrium with its environment. If the composition of the climax community is determined by the climate it is called a **climatic climax**. In most of Britain and Ireland the climatic climax is deciduous forest. Some successions do not reach a potential climax because of interference by a biotic factor, such as grazing by deer, and the final community is called a **biotic climax**.

An example of succession from bare rock is illustrated in Figure 29. Weathering of the rock and the production of organic material by the pioneer species lead to the beginnings of soil formation; further succession results in a deeper and more nutrient-rich soil.

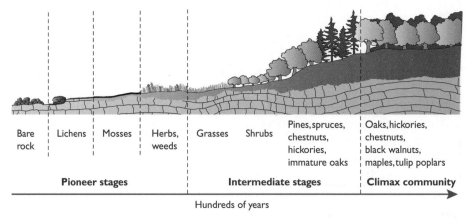

Figure 29 Ecological succession from bare rock in a North American community

As succession proceeds the following trends are evident (see Figure 30):
- The soil develops — it increases in depth and in the proportion of organic content (humus).

- The height and biomass of the vegetation increase.
- Changes in height and density of vegetation provide a greater variety of microhabitats and ecological niches.
- The increasingly complex plant community and wider variety of niches support a greater number of animal species.
- Species diversity increases, from simple communities of early succession to complex communities of late succession.
- The number of food chains increases and more complex food webs develop.
- The community becomes more stable and becomes dominated by long-lived plants (K-strategists).

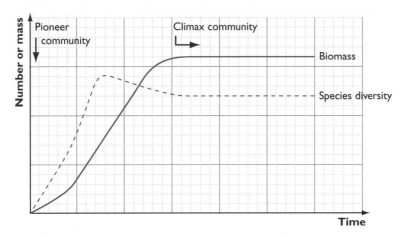

Figure 30 Some general changes in vegetation during succession

Succession takes place over long periods, e.g. hundreds of years. It is possible to 'view' succession in a sand-dune system where the zonation of vegetation over 'space' reflects changes over time. The dunes furthest from the sea were those formed first, i.e. they are the oldest; those nearest the shore formed last, i.e. they are the youngest.

Secondary succession

Secondary succession occurs after an existing ecosystem has been disturbed, e.g. by fire or flooding. Secondary succession happens more quickly than primary succession because the former type of event does not involve loss of soil and because some vegetative remains from the previous community are present. Table 14 shows the succession that would take place in an abandoned ploughed field.

Table 14 Secondary succession in an abandoned ploughed field

Pioneer	Successive seres			Climax
Annual plants (live only for one year but produce vast numbers of easily dispersed seed)	Grasses and low-growing perennials (plants which live for many years)	Shrubs and small trees	Young broadleaved woodland trees	Mature woodland, mainly oak

Examiner tip

You should be clear about the difference between *succession* and *zonation*. Succession refers to the change in the composition of a community with time. Zonation is where different communities are present at the same time, distributed as bands or zones along an environmental gradient.

Knowledge check 50

What is the difference between primary and secondary succession?

Synoptic links

You need to have an understanding of the following topics in AS Unit 2 (refer to AS Unit 2 Student Unit Guide for revision):

- the adaptations of organisms to their environment
- the ecological factors that have an influence on the distribution of organisms
- the use of quadrats, randomly positioned, to estimate plant abundance

The concept of the ecological niche should be reviewed. This is defined as 'the functional position of an organism in its environment' — comprising its habitat, the resources (e.g. for a plant, light and ion availability) it obtains there and when it obtains them, noting that resource availability will be affected by competitors (e.g. for light) while survivability will be affected by the presence of predators (or, for plants, grazing animals). An important principle is that no two species can share the same ecological niche within one habitat. For example, the two-spot ladybird (*Adalia bipunctata*) and seven-spot ladybird (*Coccinella septempunctata*) both feed on aphids. However, the two-spot ladybird is found on a narrower range of plants, where it feeds on smaller aphids, and it is not attacked by the parasitic wasp *Dinocampus coccinellae*, to which the seven-spot ladybird is susceptible. Thus the two species do not share the same niche.

Summary

- A community is all the organisms present in a particular habitat at a particular time, while an ecosystem is a community of organisms that interact with each other and with the abiotic environment.

- Succession is the progressive change in the composition of a community over a period of time: the first organisms to colonise a habitat are pioneers; colonising organisms bring about changes that allow other organisms to colonise the habitat; a series of additions and replacement of species occurs; each distinct community in a succession is called a seral stage, and the final stable community is called the climax.

- If the area has never been colonised before (e.g. a newly formed volcanic island), the process is called primary succession. If the area was previously covered with vegetation (e.g. a forest site that suffered a severe fire), secondary succession occurs.

- There are different types of climax community: the composition of a climatic climax is dependent on climate, such as temperature and rainfall; the composition of a biotic climax community is influenced by biotic factors, such as grazing.

Ecological energetics

Feeding relationships

Feeding involves the transfer of energy and materials from one organism to another. The sequence of organisms, with each being a source of food for the next, is called a **food chain**. A **grazing food chain** starts with living plants, fed on by herbivores, in turn fed on by carnivores, and so on, e.g. grass → rabbit → buzzard. A **detritus food chain** starts with dead organic matter (dead plants and animals) fed on by detritivores and decomposers. For example, dead grass → woodlice (detritivores) or fungi (decomposers). Since an organism usually feeds on several types of organism and, in turn, is fed on by more than one type, the result is a **food web**. For example, rabbits feed on herbaceous plants other than grass, while foxes and stoats, as well as buzzards, feed on rabbits.

Within a food chain or food web each organism occupies its own feeding position or **trophic level**:
- **Producers** — the autotrophic plants produce food through photosynthesis and ultimately support all other levels.
- **Consumers** — primary consumers (herbivores) feed on producers, secondary consumers (carnivores) feed on primary consumers, tertiary consumers ('top' consumers) feed on secondary consumers, and so on.
- **Decomposers** — decomposers (most bacteria and fungi) and detritivores (e.g. earthworms and woodlice) feed on the accumulated debris of dead organic matter (detritus).

Knowledge check 51

In a semi-desert ecosystem, seeds are eaten by mice, which are in turn eaten by snakes. Hawks feed on mice and snakes. What can you infer about the trophic level(s) occupied by hawks in the ecosystem?

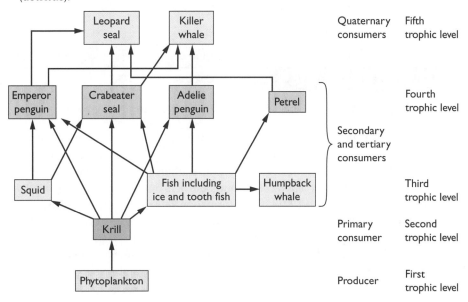

Figure 31 A food web in a marine ecosystem

Changes in any population in a food web can influence other populations. For example, in the marine ecosystem shown in Figure 31, a decrease in the number of crabeater seals will have a number of effects: it may mean that leopard seals feed more regularly on emperor penguins, which might decrease in numbers; it might mean more fish for adelie penguins, which might increase in numbers.

Food chains commonly have four links and rarely more than five. The reason for this is that not all the energy of one trophic level is available to the next. There is an 'inefficiency' of energy transfer and at the fifth trophic level there is too little energy to support a further level.

Pyramid relationships

Food chains and food webs only illustrate the types of organism. Ecological pyramids show the quantitative relationship of the trophic levels. There are three forms, each with advantages and disadvantages.

A **pyramid of numbers** represents the total numbers of the organisms at each trophic level in an ecosystem. The data for constructing a pyramid of numbers are readily obtained: simply count the organisms in each trophic level within a specified area. However, there are disadvantages:

- With some organisms, it is difficult to determine what represents a single individual, e.g. with buttercups producing runners so that several are connected.
- The approach does not take into account the size of an organism, e.g. an oak tree has more producer material than one buttercup, while a single rabbit will have lots of mites.

While pyramid relationships are frequently illustrated, inverted pyramids are not uncommon (see Figure 32).

Figure 32 Pyramid of numbers for (a) grassland and (b) an oak tree

A **pyramid of biomass** represents the total biomass of the organisms at each trophic level in an ecosystem. The data for constructing a pyramid of biomass are obtained by weighing the dry mass of the organisms at the different trophic levels. While this requires more effort than for a pyramid of numbers, it is more representative of all the material in each trophic level. Still, the information represents the **standing crop**, i.e. what is present at one moment in time and not what is being produced over time. As a result inverted pyramids are possible, especially for aquatic ecosystems where the amount of the phytoplankton (producers) at any one moment in time may seem relatively small but is being generated exceedingly quickly and so supports a larger biomass of zooplankton (consumer level; see Figure 33).

Figure 33 Pyramids of biomass for (a) woodland and (b) a marine ecosystem

A **pyramid of productivity** (also called a **pyramid of energy**) represents the energy value of new material produced at each trophic level over time. It is more difficult to obtain the data simply because measurements need to be made over a time period — measurement may be presented as, for example, $kJ\,m^{-2}\,y^{-1}$ (kilojoules per square metre per year). However, comparing the *rate* at which the organisms at each trophic level produce new material will always produce the meaningful pyramid relationship (see Figure 34).

(a)

Tertiary consumers,
e.g. sparrowhawks

Secondary consumers,
e.g. insectivorous birds

Primary consumers,
e.g. insects

Producers, e.g. oak trees

(b)

Large fish

Small fish

Zooplankton

Phytoplankton

Figure 34 Pyramids of productivity (energy) for (a) woodland and (b) a marine ecosystem

Productivity is the energy entering a trophic level that remains as energy in biomass:
- for a given area of the ecosystem (often a square metre)
- in a given period of time (often a year)

So productivity is the *rate* of biomass production (units, for example, $kg\,m^{-2}\,y^{-1}$) or of energy accumulation (units, for example, $kJ\,m^{-2}\,y^{-1}$).

The inefficiency of energy transfer in ecosystems

There are a number of reasons for the inefficiency of energy transfer and the precise nature of each of these depends on the trophic level.

Reasons for the low percentage of solar radiation absorbed by plants in photosynthesis: Of the total amount of solar energy arriving at the Earth only a small fraction is used by the producers (i.e. in photosynthesis). At the equator during the day, about 1.4 kJ of energy every second reaches the upper atmosphere from the sun over every square metre. Most of this solar energy *never reaches the ground* since over 99% is:
- *reflected back* into space by clouds and dust, or
- *absorbed by the atmosphere* and re-radiated

Of the solar energy that reaches the ground most will not be used since:
- most will *miss the leaves altogether*, with less than 0.1% actually reaching the surface of the leaves

When light from the sun does hit leaves, a mere 0.5 to 1% of the incident energy is actually used by a plant in photosynthesis since:
- some of the energy is reflected by the leaf, or
- some is transmitted through the leaf and misses the chlorophyll molecules, or
- over half of the light that strikes the chlorophyll consists of wavelengths which cannot be used in photosynthesis (such as green or ultraviolet light), or
- the reactions of photosynthesis themselves are inefficient, losing much energy as heat

Primary productivity: The 0.5–1% of incident light energy that is converted into chemical energy and fixed by producers in photosynthesis is called the **gross primary productivity** (**GPP**). Some of the GPP is required by the plant for **respiration** (e.g. generating ATP for the active uptake of ions). Subtracting respiration from GPP gives the **net primary productivity** (**NPP**). The NPP is the actual rate of production by the producers (autotrophic plants), and is important because it represents the energy or biomass available:
- for the new growth of the plants, and
- to all the other trophic levels in the ecosystem

Examiner tip

You should be aware that productivity of ecosystems has two components: **primary productivity** — the production of new organic matter by plants; and **secondary productivity** — the production of new organic matter by consumers. You must be able to distinguish between the two.

Examiner tip

You may be asked about reasons for the inefficiency of light absorption by plants. Make sure you read the question carefully. If the question requires reasons for the inefficiency of light absorption *once light has reached the vegetation*, then 'absorption by the atmosphere' is not a valid answer.

Examiner tip

You must know that the initials GPP stand for 'gross primary productivity', while NPP represents 'net primary productivity'; and you must be able to distinguish between them (NPP = GPP − respiration).

Knowledge check 55

Why is the net primary productivity (NPP) often used to compare the efficiencies of different ecosystems?

Reasons for the reduction in energy at progressive trophic levels: When energy is transferred from producers to primary consumers, and to each trophic level thereafter, the efficiency is only about 5–20%. This is because:

- some of the material is *not consumed*, either because it is inaccessible or is unpalatable or inedible (e.g. plant roots cannot be eaten by grazers, some plants may be spoiled by animal droppings, the hooves of a wildebeest are not eaten by a lion, the shell of a snail is not eaten by a song thrush)
- some of the material is *not digested*, so is not absorbed and appears in the faeces (e.g. skin and bones in the droppings of a barn owl, cellulose is not easily digested and so is egested in the faeces of herbivores)
- some of the material ends up as a waste product of metabolism and is *excreted* (e.g. urea in urine)
- many of the materials are *used in respiration* to generate ATP for active processes in the organism (e.g. ATP used in muscle contraction)

Secondary productivity: For animals, **energy budgets** can be constructed from the quantities of energy consumed (C), remaining as the **net secondary productivity** (**NP**), released in respiration (R) and leaving the animal as urine (U) or faeces (F). The equation for NP (in units of energy per unit time, e.g. $kJ\,y^{-1}$) is:

$$NP = C - (R + U + F)$$

For livestock (e.g. cattle), it is the net secondary productivity, NP, that represents the energy available for human consumption.

The transfer of energy through an ecosystem and the efficiency of this transfer between trophic levels is shown in Figure 35.

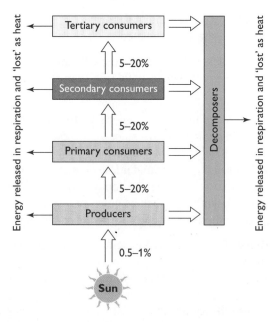

Figure 35 The transfer of energy through an ecosystem

The range of 5–20% for the efficiency of energy transfer through the consumer trophic levels is a guide only and, indeed, is often summarised simply as 'the 10% rule'. The

---

Knowledge check 56

The amount of energy consumed (C) less the amount egested as faeces (F) represents the amount of energy absorbed (A). Rewrite the equation for net secondary productivity (NP) to *include* the amount of energy absorbed (A).

Knowledge check 57

Arctic ecosystems most often have only three trophic levels, while tropical ecosystems frequently have five levels. Explain why.

CCEA A2 Biology

actual figure for the percentage efficiency of energy transfer will depend on the nature of the organisms, for example:

- herbivores (primary consumers) generally have a low percentage efficiency, e.g. 13% for a grasshopper (herbivore) and 27% for a wolf spider (carnivore), because of the difficulty in digesting the cellulose in plant material, so there are relatively high losses via the egestion of faeces
- endotherms (mammals and birds) have a low percentage efficiency (e.g. mammals have a figure of only 1–4%), because they generate heat internally to maintain a relatively high body temperature, which means their metabolism is maintained at a high rate but there are relatively high energy losses through respiration

The percentage efficiency of energy transfer between trophic levels is calculated as:

$$\text{efficiency of energy transfer} = \frac{\text{energy in one trophic level}}{\text{energy in previous trophic level}} \times 100\%$$

Productivity has been determined for an entire ecosystem — Silver Springs in Florida. The energy in gross production, net production after respiration and the energy going to the decomposers for this ecosystem is shown in Figure 36 (the unit of measurement is $kJ\,m^{-2}\,y^{-1}$).

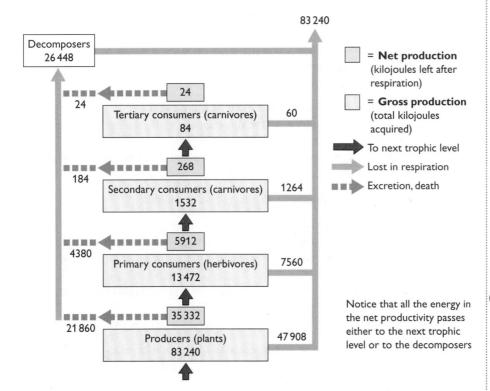

Figure 36 Gross and net productivities in the Silver Springs ecosystem

Notice that all the energy in the net productivity passes either to the next trophic level or to the decomposers

Look at the figures carefully. It is evident that there is less energy at each successive trophic level, but note other important findings:

- *more energy is used in respiration than is available as net production* for the next trophic level — respiration generates energy for all the active processes requiring ATP in the organisms and the energy is then 'lost' as heat
- *more of the energy from net production passes to the decomposers* than to the next trophic level of the grazing food chain — for example, in a field of grass, some of the grass will have been trampled by large grazers or will have been spoiled by animal faeces and so is only available to the decomposers, while leaves may simply die

Some of the energy in an ecosystem may be lost by the 'emigration' of material out of the ecosystem and be gained by the 'immigration' of material into it.

The implications for agriculture

The loss of energy in food webs has important implications when it comes to the production of food and our position as consumers.

Since more energy is available to primary consumers than to secondary consumers, more energy will be available to humans eating plant material than if animal products are eaten. A vegetarian diet will support many more people than a diet of meat, and in many heavily populated developing countries meat is seldom eaten. However, this does not mean that all animal husbandry is wasteful of resources. For example, in the upland areas of Northern Ireland the soil and climate will not support a crop plant which humans can eat. In these areas it makes ecological sense to raise animals, such as sheep, which can tolerate the poor-quality vegetation growing there, and then to eat these animals and/or their dairy products.

Where plant crops are grown productivity can be increased by:
- use of fertilisers, so that the plants have more ions, e.g. the nitrogen (N) in nitrates is used in the production of proteins
- use of pesticides, e.g. herbicides will prevent competition from non-crop plants, while insecticides will reduce damage from insect pests

There are concerns about overuse of fertilisers and pesticides.

Where animals are farmed productivity can be increased by:
- feeding the animals on high-protein foods and high-energy foods such as silage (silage is made by cutting grass and enclosing it in plastic, where it is fermented by bacteria)
- keeping animals, such as chickens and pigs, in warm conditions, so that less energy is used to generate heat
- keeping animals in confined conditions, so that less energy is used in movement

There are ethical issues here, and some people think that it is cruel to keep animals in crowded and confined conditions where movement is restricted; some consider that as long as the animals are kept comfortable with sufficient food and water then the benefit of producing food cheaply outweighs any other concern.

Knowledge check 60

Explain why a vegetarian diet can support more people.

- Light energy is trapped by producers (plants) in photosynthesis and is the external energy source for most ecosystems. Only a very small percentage of the total light energy available is utilised by producers.
- Energy is passed through food chains, which are linked with other chains as food webs.
- Energy is passed from producers to primary consumers (herbivores), to secondary consumers (carnivores), to tertiary and, perhaps, to quaternary consumers. Each level in the food chain is called a trophic level.
- Decomposers (mostly bacteria and fungi) break down the dead organic matter from producers and consumers, and play an important role in nutrient recycling.
- There are three types of ecological pyramid: pyramids of numbers, pyramids of biomass and pyramids of energy. Pyramids of energy are the most useful and show that approximately 90% of the energy transferred to the next trophic level is lost as waste matter or heat.

- Transfer of energy to carnivores is more efficient than to herbivores; endotherms (mammals and birds) have a very low efficiency of energy transfer.
- Productivity is a measure of how efficient organisms are at locking up energy in organic molecules in their body cells.
- Gross primary productivity (GPP) is all the biomass produced (or energy 'trapped') by plants (producers) per square metre per year. Since some energy is used in respiration, this leaves the plants with a net primary productivity (NPP).
  GPP = NPP + respiration
- Secondary productivity is the rate of production of biomass (or gain in energy 'trapped') by animals (consumers) in an ecosystem.
  net secondary productivity = amount consumed − amount lost in faeces − amount lost in urine − amount used in respiration

Nutrient cycling

Energy is *transferred through* ecosystems; it is not recycled. This is simply because in all energy conversions heat energy is released and this eventually radiates out into space. However, the elements in matter are recycled. Each element spends part of its time in complex organic molecules and part in simple, inorganic substances in the abiotic part of an ecosystem.

The carbon cycle

Plants (producers) absorb carbon dioxide (CO_2) *from the atmosphere* (or as hydrogen carbonate ion, HCO_3^-, from water) and synthesise carbohydrates during **photosynthesis**. From the carbohydrates produced, and using atoms in absorbed ions, the autotrophic plants produce lipids, proteins and nucleic acids, all of which contain carbon. These molecules are **consumed** by the animals (consumers) in the food chain, entering their bodies following digestion.

Both the plants and the animals release carbon dioxide via their **respiration**.

Decomposers (most bacteria and fungi) use dead plants and animals for food. Some of this is used in **respiration**, releasing carbon dioxide.

Fossil fuels such as oil, coal, peat and gas contain carbon. When they are **burned**, carbon dioxide is released back into the atmosphere.

Examiner tip

Remember energy in an ecosystem is not recycled. There must be an external energy source (usually sunlight) since energy is lost as heat as it passes along food chains. On the other hand there is a finite amount of matter available, and so elements that make up organic organisms must be recycled when they die.

Knowledge check 61

By what process does carbon in the atmosphere become available to organisms in a community?

These processes are summarised in Figure 37.

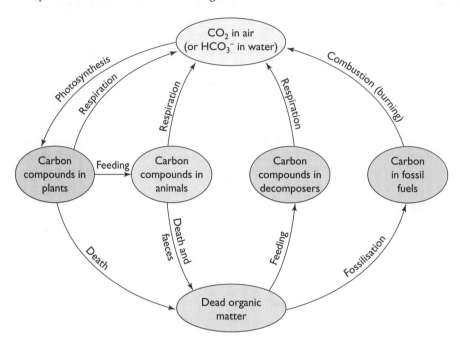

Figure 37 The carbon cycle

The nitrogen cycle

Plants (producers) absorb nitrate ions (NO_3^-) from the soil (or water, if the plants are aquatic). The plants use the nitrogen in these ions to *synthesise amino acids, and so proteins, and nucleotides, and so nucleic acids*. These molecules are consumed by the animals (consumers) in the food chain, and enter their bodies following digestion. Undigested matter is egested as faeces.

Animals produce nitrogenous waste products, ammonia and urea, which are excreted.

Decomposers (decomposing bacteria and fungi) use dead plants and animals, faeces and urea for food. The protein (and their amino acids) and nucleic acid (and their nucleotides) are broken down and the nitrogen *released as ammonia* (or ammonium ion, NH_4^+).

The ammonia (NH_3) is used by **nitrifying bacteria** and *converted (oxidised) to nitrate ion* (NO_3^-). Nitrates are released when the bacteria die.

This description covers the basis of the nitrogen cycle (see Figure 38), except that other processes occur to either add to or reduce the nitrogen available to plants:
- **nitrogen-fixing bacteria** *add to the nitrogen available to plants* — nitrogen-fixing bacteria, some species of which (e.g. *Rhizobium*) inhabit the root nodules of legumes (e.g. clover and beans), are able to utilise gaseous nitrogen to synthesise organic nitrogen-containing compounds (i.e. amino acids)
- **denitrifying bacteria** *reduce the nitrogen available to plants* — these bacteria, living in waterlogged and oxygen-deficient soils, use nitrate and convert it to gaseous nitrogen which is returned to the atmosphere

Examiner tip

Many students find the nitrogen cycle difficult. You should appreciate that nitrogen is an element which is a constituent of some inorganic ions and some organic molecules: ammonium (NH_4^+) and nitrate (NO_3^-) ions are inorganic; amino acids (and so proteins) and nucleotides (and so nucleic acids) are organic. The term 'nitrogen' may be used in questions to cover the total nitrogen in an ecosystem, whether in inorganic or organic form.

Knowledge check 62

In what form do plants absorb nitrogen from their environment?

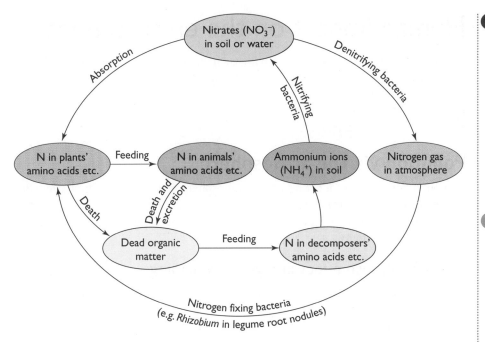

Figure 38 The nitrogen cycle

Knowledge check 63

By what processes do some bacteria convert: (a) organic matter into ammonia; (b) gaseous nitrogen into ammonia and amino acids; (c) ammonia into nitrates; (d) nitrates into gaseous nitrogen?

Examiner tip

You may be asked to reproduce or complete a diagram of the nitrogen cycle in your exam. Study Figure 38 and copy it until you can draw the cycle without reference to the figure.

Knowledge check 64

Which of the bacteria involved in the nitrogen cycle respire anaerobically and so favour waterlogged soils?

In farm ecosystems, removing crops or grass (for silage) removes some of the primary production so that less is available to any grazing food chain. It also means that nitrogen-containing compounds within the plants are removed so that the soil becomes depleted in nitrate unless this is added in organic or inorganic fertilisers.

Synoptic links

You need to have an outline understanding of the structure of carbohydrates, lipids, proteins, nucleic acids and ATP. In particular, you should recognise that proteins, nucleic acids and ATP are nitrogen-containing compounds. Refer to the AS Unit 1 Student Unit Guide for revision.

Summary

- Carbon is found in all organic molecules.
- In the carbon cycle:
  - CO_2 is removed from the air and fixed into organic molecules in photosynthesis
  - CO_2 is replaced in the air when any organism respires or when fossil fuels are burned
  - organic molecules are passed from plants to animals by feeding and assimilation
  - organic molecules are passed from plants and animals to decomposers, which feed on the dead remains of organisms
- Nitrogen is found in amino acids (proteins), nucleotides (nucleic acids) and ATP.
- In the nitrogen cycle:

  - nitrate (NO_3^-) ions are absorbed from the soil by plants and used to make amino acids, proteins, etc.
  - animals eat plants, digest the plant protein and use their amino acids to make animal protein
  - when the plants and animals die, decomposers release ammonium (NH_4^+) ions from nitrogen-containing molecules
  - nitrifying bacteria oxidise the ammonium ions to nitrates
  - nitrogen-fixing bacteria (free in the soil or in root nodules of legumes) reduce nitrogen gas to ammonium ions and hence to amino acids
  - denitrifying bacteria use nitrate and release N_2 to the atmosphere

Human impact on the environment

All organisms exchange materials with their environment. However, the impact that humans have on the environment is unlike that of any other species. Areas of particular concern are:
- pollution via the atmosphere and of waterways
- agriculture and forestry

Pollution via the atmosphere

Air pollution is the introduction into the atmosphere of substances that harm living organisms or damage the environment. Some of these substances occur naturally but are pollutants because of the damagingly high concentrations generated as a result of human activity.

The greenhouse effect and global warming

The greenhouse effect is a natural process that occurs all the time. It keeps the Earth's surface warm; without it, the temperature would be about –18°C. Energy from the sun reaching the atmosphere is absorbed, reflected back into space, or transmitted to the Earth's surface. Energy reaching the Earth's surface is radiated back into space at longer wavelengths than the energy arriving from the sun. Some of this longer-wavelength energy (heat) is absorbed by various gases in the atmosphere, and re-radiated back towards the Earth's surface. These gases are called **greenhouse gases** and the warming that this produces is the **greenhouse effect** (see Figure 39).

Examiner tip

You should be aware that the greenhouse effect is a natural phenomenon that keeps the Earth warm. This is why the present problem arising from *increased levels* of greenhouse gases is better known as the *enhanced greenhouse effect*.

Knowledge check 65

Explain the rise of carbon dioxide levels in the atmosphere.

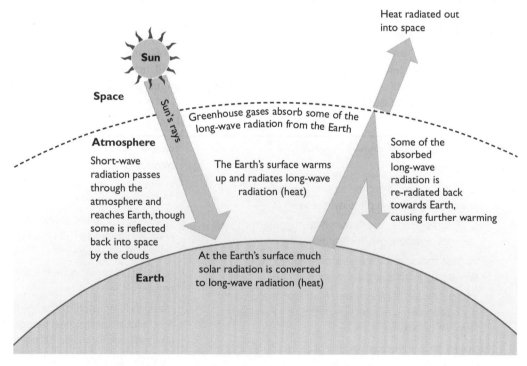

Figure 39 The greenhouse effect

The problem arises from the *increase in the levels of these gases* in the atmosphere as a result of human activities. Information about these gases is given in Table 15.

Table 15 Greenhouse gases, their sources and other information

Name of gas	Main sources	Comments	Remedies
Carbon dioxide (CO_2)	• respiration of organisms • combustion of fossil fuels • deforestation so that less CO_2 is taken up in photosynthesis	About 50% of the greenhouse effect (ignoring the effect of water vapour; see *Examiner tip* opposite) is thought to be caused by CO_2 from burning fossil fuels	1 Reduce the use of fossil fuels, i.e. use other energy sources and preferably renewable sources 2 Plant new forests to act as carbon sinks
Methane (CH_4)	• bacteria living in marshes and rice paddy fields • bacteria in the guts of cows and other ruminant animals • decomposition of organic material in landfill sites	A molecule of methane is over 20 times more potent as a greenhouse gas than CO_2, though since there is so much more CO_2 the latter has the greater overall effect	Use digesters to decompose organic waste and burn the methane produced, e.g. energy obtained from the combustion of methane from digesters in sewage works can be used to run the works
Nitrous oxide (N_2O)	• denitrifying bacteria acting on nitrates • combustion of fossil fuels	The increased use of artificial nitrate fertilisers has led to more N_2O being produced	1 Reduce the use of artificial fertilisers (especially on soils liable to waterlogging) 2 Reduce use of fossil fuels
Chlorofluoro-carbons (CFCs)	• coolant in refrigerators • propellants in aerosol sprays • in the packaging material Styrofoam (CFCs released on incineration)	CFCs are very potent greenhouse gases — 1200 times more effective than CO_2; CFCs are entirely human-made	Use alternatives to CFCs
Ozone (O_3)	• produced by action of sunlight on gases from vehicle exhausts	O_3 as a greenhouse gas operates in the lower atmosphere — not to be confused with the 'ozone layer' in the stratosphere	Reduce vehicle exhaust emissions

Thus, there is an increase in the greenhouse effect (**enhanced greenhouse effect**), producing steadily increasing temperatures on the Earth's surface, a condition called **global warming**. The possible effects of global warming are:
- **a rise in sea levels** — resulting from the melting of polar ice caps and glaciers and thermal expansion of water, which would cause flooding of low-lying areas and coastal erosion
- **disruption of climate and weather patterns** — resulting from a change in wind patterns and leading to changes in the distribution of rainfall and greater extremes in the weather

Examiner tip

Water, in gaseous form (water vapour) and in liquid form (tiny droplets in clouds) is the *major* greenhouse gas, contributing approximately 75% of the greenhouse effect. However, about 90% of the water in the atmosphere is the result of evaporation (mostly from oceans), while most of the remaining 10% comes from evapotranspiration. These are not influenced significantly by human activity and so water as a greenhouse gas contributing to the *enhanced* greenhouse effect is omitted in Table 15.

Examiner tip

Greenhouse gases work because their molecules absorb long-wave radiation (infrared radiation or heat). Interestingly, gaseous nitrogen (N_2) and oxygen (O_2), which together make up 98% of the atmosphere, are incapable of absorbing long-wave radiation. All greenhouse gases consist of three atoms or more.

- **changes in crop production** — resulting from a gradual increase in temperature at the Earth's surface and possibly including massive reductions in grain crop yields in North America
- **altered distribution of wild animals and plants** — resulting from temperature changes, e.g. there is evidence that the extent of the sea ice at the North Pole is reducing and, since this is the habitat of the polar bear, it is a threat to their future success; in Britain there is evidence that the range of the speckled wood butterfly is gradually spreading northwards; 34 of the 37 British species of dragonfly have expanded their range northwards; the starfish, *Solaster endeca*, may be an indicator of global warming in Northern Ireland where it is at the southern limit of its shallow-water distribution

Global warming is a controversial issue. An issue is controversial when alternative points of view about it can reasonably be held. Global warming remains controversial because:

- Science cannot prove theories — the scientific method can only disprove them. With the scientific method, a hypothesis is proposed to explain an observation, and then the hypothesis is tested. If the results disprove the hypothesis it is rejected, whereas if the results support the hypothesis it does not actually prove it — there could be alternative explanations.
- Simply establishing a correlation between carbon dioxide levels and an increase in global temperature does not mean that there is necessarily a causal relationship; and even if there was a causal relationship it does not mean that CO_2 increases cause global warming — it could be that global warming is stimulating rates of respiration and so causing an increase in CO_2 levels, as suggested by at least one prominent biologist.
- There is incomplete knowledge of how the climate systems of our planet work, and the datasets used in making predictions about climate change have their limitations. For example, there is no way to measure precisely how much carbon dioxide is added to the atmosphere by fossil fuel combustion or by respiration, particularly from decomposers.

Despite much controversy, there is widespread scientific consensus that temperatures are indeed rising, and that rising levels of greenhouse gases are at least partly responsible. This consensus results from the gradual build-up of a large body of scientific evidence supporting the theory. There are, nevertheless, alternative opinions.

Ozone depletion

The **ozone layer** is present 10–45 km up in the stratosphere. It is formed by the effects of **ultraviolet radiation** on oxygen. The ozone layer acts as a screen, preventing much of the damaging UV radiation from reaching the Earth's surface.

Chlorofluorocarbons (**CFCs**) are known to break down ozone (O_3). They have been artificially produced for the functions shown in Table 15. While CFCs have been phased out in most countries of the world, their effects continue. CFC molecules can stay in the stratosphere for about 100 years, and each one can break down hundreds of thousands of molecules of ozone.

The ozone layer has been thinned to as little as 67% of its former density in places and 'holes' have been identified over the poles, especially in Antarctica. The effect of ozone depletion is to allow more of the harmful UV radiation to reach the Earth's surface. UV radiation is known to be associated with an increase in incidence of **skin cancers** (melanoma) and **cataracts** (clouding in the lens of the eye). Using blocking creams and wearing a wide-brimmed hat reduces the risks of skin cancer. Wearing sunglasses gives additional eye-protection from UV radiation.

Acid rain and acidification of ecosystems

Rainwater is naturally slightly acidic, with a pH between 5.5 and 6.5, since the atmosphere contains carbon dioxide which dissolves in water to form carbonic acid. The term **acid rain** is used to describe rain (or snow) with a pH of less than 5. The main causes of acid rain are emissions of **sulfur dioxide** (SO_2), **nitric oxide** (NO) and **nitrogen dioxide** (NO_2). Sulfur dioxide is produced from the burning of coal and oil, mainly by power stations, while nitrogen dioxide is produced in internal combustion engines and so is given off in vehicle exhaust emissions. These gases dissolve in water to form sulfuric acid and nitric acid. These acids can be carried within clouds for long distances before coming down as acid rain (so that one country can suffer pollution generated in another).

Forest and aquatic ecosystems appear to be particularly susceptible to the *effects* of acid rain:

- *in forest ecosystems*, acid rain is *directly toxic to leaves* and pine needles and contributes to defoliation
- acid rain *inhibits the action of decomposers* in the soil, and so the cycling of nutrients
- acid rain *influences the solubility of mineral ions* in the soil — such as calcium, magnesium and potassium — which are needed as plant nutrients
- in very acid conditions, *aluminium becomes soluble* and is released into soil water where it is *toxic to roots*
- *in aquatic ecosystems, aluminium leaches into waterways and causes gills to become covered in a thick mucus which prevents oxygen uptake* and so kills invertebrates (e.g. mayfly nymphs) and fish
- *fish eggs fail to hatch* if exposed to pH levels below 5.5

A number of *measures* have been put in place to tackle the cause of the problem:

- using low-sulfur fuels
- increasing the use of natural gas in place of coal which has a high sulfur content
- removing sulfur dioxide from waste gases produced by burning coal in power stations (i.e. flue gas desulfurisation using 'scrubbers' or filters)
- using catalytic converters in car engines, oil burners etc., to remove nitrogen dioxide

The pollution of waterways

Lakes and rivers are polluted by the addition of organic matter or inorganic nutrients, since these can adversely affect human health and the survival of other organisms.

Organic pollution

Organic pollution includes sewage, slurry from intensive livestock units, silage effluent, washings from dairies and papermill waste products.

Examiner tip

Many students confuse the different effects of atmospheric pollution: greenhouse effect (global warming), ozone depletion and acid rain. There is little excuse for this – they are quite distinct. Nevertheless, you should be aware that some gases may have duel effects — for example, CFCS are greenhouse gases though they are a more important agent in ozone depletion.

There are bacteria in water which use the organic waste as food. These decomposing bacteria use up oxygen as they respire, reducing the oxygen level in the water. They place a high oxygen demand upon the water. The quality of a body of water can be measured by its **biological oxygen demand** (**BOD**). The *lower the BOD*, the fewer bacteria are present, indicating that there is *less organic material* in the water.

Whenever organic waste such as sewage gets into a river, there are changes downstream from the point of discharge. Not only are there changes in the BOD and oxygen levels (see Figure 40a), but there is also a succession of different species of aquatic invertebrates (see Figure 40b). Specific species are adapted to different levels of oxygen in the river, and thus act as '**indicator species**' of the degree of pollution.

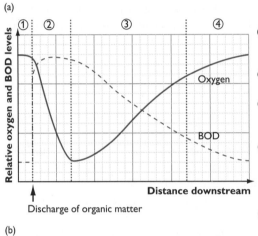

(a)

① Initially, the discharge of organic matter provides food for bacteria, which multiply dramatically — there is a high BOD

② The bacteria use up oxygen so that oxygen levels fall

③ As the organic matter is decomposed the oxygen levels start to rise and the BOD falls

④ Eventually, much of the organic matter is used and the oxygen levels return to the normal level

① Clean-water invertebrates, such as stonefly larvae and mayfly nymphs, and fish, such as trout, swim away or are killed since they cannot survive in water with little oxygen

② Only organisms adapted to survive in water with very low levels of oxygen, such as sludge worms (*Tubifex*), can survive, so their numbers increase

③ As oxygen levels gradually rise there is a succession of organisms adapted to lower-than-normal oxygen levels, such as midge larvae (*Chironomus*) and waterlice (*Asellus*)

④ As oxygen levels return to normal, pollution-tolerant species decline due to increased predation and competition from returning pollution-intolerant (clean-water) species

Figure 40 The discharge of organic matter into a river causes changes in (a) oxygen levels and BOD, and (b) the distribution of organisms in the river

Discharge of organic matter into waterways is regulated. Where it is unavoidable (e.g. from septic tanks, or in small amounts from farms), reed beds may be constructed at the point of discharge. The common reed (*Phragmites australis*) has the ability to transfer oxygen from its leaves, down through its stem, and out via its root system. The high levels of oxygen released support a large bacterial population capable of digesting organic matter in the discharge, as long as the latter is not excessive.

Eutrophication

Eutrophication is the naturally occurring process of nutrient enrichment of water bodies as a result of rock erosion and run-off from the surrounding catchment. However, artificial enrichment is occurring as a result of human activities, changing the biological communities of lakes, ponds and some rivers. Lough Neagh and Lough Erne in Northern Ireland are eutrophic lakes.

Eutrophication is accelerated by the following:
- increased leaching and run-off of nitrate-rich and phosphate-rich artificial fertilisers from agricultural land
- release of phosphate-containing detergents in household waste
- products of the decomposition of organic matter discharged into waterways (see above)

Eutrophication may have the following *consequences*:
- increased nitrate and phosphate loading causes massive increases in amounts of algae, particularly blue-green algae (cyanobacteria), i.e. **algal blooms**
- dense algal blooms cut down light penetration and use up available ions so that *algae die*, as they do not receive sufficient light and/or nutrients
- dead algae are decomposed by *bacteria* which *use up oxygen* as they respire
- oxygen depletion means that *many species of invertebrates and fish die*

Possible adverse conditions created include:
- water extracted for drinking having an unacceptable taste or odour
- water being harmful to health, as it may be contaminated by toxic blue–green algae, while high nitrate levels may be toxic
- rooted plants dying (due to insufficient light) and food chains relying on these collapsing, so biodiversity is reduced
- important fisheries being lost

Remedies have been applied to reduce the risk of eutrophication.

First, to reduce the risk of fertiliser run-off into waterways:
- testing soils for ion content so that the amount of any fertiliser applied has been calculated to ensure that 'supply does not exceed demand'
- only applying artificial fertilisers immediately prior to or during periods of vigorous plant growth
- not using artificial fertiliser when heavy rain is forecast
- not applying fertilisers adjacent to water bodies

Second, we can remove phosphates (considered to be the most limiting factor with respect to blue-green algal populations) from sewage works' effluent, e.g. all the sewage works around Lough Neagh have this feature.

Agriculture

The human activity that most impacts on the environment is agriculture.

Knowledge check 66

Suggest one reason why the consequences of eutrophication are: (a) more likely to occur in warm water; (b) less likely to occur in moving water than in still water.

Intensive farming

Traditional (**extensive**) **farming methods** operate with a minimum of input from outside the farm. The farm supports a wide variety of crops (e.g root crops, cereals) and mixed livestock (some sheep, cattle, pigs and chickens). **Crop rotation** is used to reduce the risk of mineral ion deficiency in the soil and the threat to the crop from pests. Manure from the livestock is used to fertilise the crops and some of the crop is used to feed the livestock. The produce is used locally.

Intensive farming methods were developed during the 1950s to *increase yields*. High yields required high inputs and benefited from specialisation; and specialisation meant the growth of **monocultures** (single-species crops). However, too many problems result from intensive farming, so it is not sustainable. Features associated with intensive farming and the ensuing problems are presented in Table 16.

Table 16 Features associated with intensive farming and ensuing problems

Features associated with intensive farming	Problems
Concentration on a small number of crop plants which are grown in large areas of **monoculture**	• Depletes the soil of particular mineral ions • Allows pests and diseases to build up year after year (a pest, in agriculture, is any organism which causes a loss in crop yields exceeding 5–10% of total yield)

Use of a **wide variety of pesticides**:
• **herbicides** to kill competing weeds
• **insecticides** to kill insects feeding on crop plants
• **fungicides** to kill species of parasitic fungi infecting crop plants
	• Pests are r-selected species with rapid rates of population growth, so **resistant strains tend to evolve**
	• Pesticides may be toxic to a variety of species, not just the pest species (these are '**broad-spectrum**' pesticides), leading to:
	  – **pest resurgence** if the pest returns in even greater numbers than before if a natural predator is killed
	  – **secondary pest outbreak** if a minor pest, previously kept at low numbers, multiplies rapidly in the absence of its competitor
	• Pesticides may **persist in the environment** (i.e. remain active over many years) and adversely affect other ecosystems: e.g. insecticides may enter the soil ecosystem, where some insects may be important detritivores, and disrupt food webs; insecticides enter aquatic ecosystems where, again, food webs are disrupted; fungicides entering the soil ecosystem would adversely affect nutrient cycling since fungi are important decomposers; herbicide spray may drift into adjacent areas and kill plant species important in the food chains of various animals
	• Persistent pesticides may also be **non-biodegradable** (not broken down in animal tissues and not excreted), so there is a build-up of the pesticide along the food chain — a phenomenon called **bioaccumulation**

Features associated with intensive farming	Problems
Removal of hedgerows to provide more growing space and make using heavy machinery easier	• **Crops lose shelter** from the effects of wind • Increased risk of **soil erosion** • **Reduced species biodiversity**, as hedges are important habitats for plants and animals
Use of **large amounts of artificial (chemical) fertilisers** to increase yield from the depleted soil	• **Loss of soil crumb structure** as no organic matter is added to the soil • Increased risk of **eutrophication** of waterways • **Reduced species biodiversity** (as low-nutrient-tolerant species are outcompeted)
Draining of marshy areas and **removal of woodland** to increase the area of land available for cultivation	Reduced species biodiversity (as other habitats are lost)
Disposal of **large quantities of slurry** produced by intensively farmed animals	• Increased risk of **organic pollution** in waterways • May contain **toxic residues of veterinary medicines**

Sustainable farming

Farming is the means by which food is produced — and staple foods need to be produced cheaply. A balance is required by which farming methods are efficient enough to provide the amount of food required for the short term, but they protect the environment to ensure food production can be maintained well into the future.

Features associated with sustainable farming and the resulting benefits are presented in Table 17.

Table 17 Features associated with sustainable farming and resulting benefits

Features	Benefits
Promotion of **polyculture** (instead of monoculture) • crop rotation, with alternation of crops from year to year, including legumes • intercropping, with two or more crops grown at the same time on the same plot, often maturing at different times	• Supports greater species diversity • Harder for pest populations to become established • Pest-control and fertiliser costs are reduced
Increase in the use of **organic fertilisers**, i.e. farmyard manure (reducing need for artificial fertilisers; where artificial fertilisers need to be used guidelines and directives are applied as described under 'remedies' in the section on eutrophication)	• Improves existing crumb structure, and so aeration and drainage • Reduces risk of soil erosion (since soil is less 'dusty') • Reduces risk of leaching and thereby of eutrophication • Uses up accumulated farm waste, such as pig slurry
Replanting of hedgerows and improving maintenance of existing hedgerows	• Provides greater shelter for crops and livestock • Decreases risk of soil erosion • Improves species biodiversity
Ploughing across slopes (not down), creating terraces	**Reduces the risk of soil erosion** when rainwater flows down the slope

Examiner tip

Draining land and ploughing fields is beneficial on existing agricultural land because it ensures that anaerobic bacteria cannot compete with aerobic bacteria, thus preventing denitrification. However, the drainage of wetlands to create more agricultural land has resulted in a significant reduction in biodiversity (40% of known species are found in wetlands) as well as contributing to more intensive downstream flooding.

Table 17 (*continued*)

Features	Benefits
Use of biodegradable plastic around and between crop plants (e.g. maize)	**Prevents growth of weeds**, reducing dependence on herbicides
Leaving **crop stubble over the winter**, i.e. plough in later	• Reduces risk of soil erosion (less bare soil to blow away) • May provide some winter feed for birds etc. and so improve biodiversity
Use of integrated pest management (IPM) schemes	See Table 18

Knowledge check 67

Explain why organic fertilisers release nutrients over a longer period of time than do inorganic fertilisers.

Pests are dealt with using integrated pest management (Table 18). This involves the development of an overall strategy, with a range of control measures and the goal of significantly reducing or eliminating the use of pesticides while at the same time managing pest populations at an acceptable level.

Table 18 Features of integrated pest management (IPM)

IPM measures	Comments
Selecting varieties best suited for local growing condition	The most productive variety would be avoided if it was susceptible to some local pest
Crop rotation and/or intercropping	Harder for pest species to become established
Monitoring pest levels	To determine the need for control measures
Use of methods to disrupt the pest's breeding	• Use of mechanical traps • Use of sterile males of pest species
Use of specific, natural predators or parasites of the pest (biological control)	• Long-term control without the problems associated with chemical control • Emphasis on control not eradication
Use of narrow-spectrum, non-persistent (biodegradable) pesticides	• Fast response to a sudden surge in pest numbers • Does not affect other ecosystems and neither is there bioaccumulation
Selective breeding of varieties of crops (or development of genetically modified strains)	New strains developed for improved pest or disease resistance

Knowledge check 68

Explain why modern agricultural practices may involve selective breeding to produce new strains of crops.

Sustainable forestry

Woodland is an important resource, and particularly so in Ireland. Northern Ireland is the least wooded country in Europe (apart from Iceland): 8% of the land area of Ireland is under woodland compared with a European average of 38%. Sustainable forestry takes into consideration the varied roles of woodland: it is a valuable source of timber (particularly for the building sector); it provides habitats for wildlife and so is important in sustaining biodiversity; and it has an important role in recreation and leisure activities. Trees removed for timber may be replaced with young trees so that the woodland is regenerated. The following factors may be considered:

• Sitka spruce, lodgepole pine and larch, which are softwood species, are frequently planted since they have relatively high yields and so generate more timber.
• Management of timber production is more considerate and alternative systems to clear-felling (large areas of woodland cut down at the same time) are used.

- Coppicing (management based on regeneration by regrowth from the cut stumps) can be used for some hardwood (broadleaved) species, such as willow, hazel and ash, and causes least disruption to the forest ecosystem.
- Farmers have been encouraged to 'set aside' land (previously used for food production) and to plant trees.
- New plantings use a variety of species to improve the aesthetic appeal of the forested countryside.
- Ancient woodland that is left (and this accounts for less than 1% of land cover in Northern Ireland) is preserved. This ancient woodland maintains a very high level of biodiversity.
- There is an emphasis on planting more trees of native broadleaved species, such as ash and oak, since these provide a greater variety of habitats and increased biodiversity compared with introduced softwoods such as Sitka spruce.

Knowledge check 69

What is meant by the phrase *sustainable management of wood production*?

Synoptic links

You need to have an understanding of the following topics, which you should revise:
- human impact on biodiversity
- strategies required to encourage biodiversity

These topics are covered in the AS Unit 2 Student Unit Guide.

Summary

- A rapidly increasing human population, possibly because of exploitation of environmental resources and developments in agriculture, has resulted in disruption and damage to ecosystems.
- An enhanced greenhouse effect, brought about by increases particularly in carbon dioxide and methane, may be producing global warming and long-term climate change. This may bring about changes in the breeding season and habitat range of many plants and animals.
- Ozone depletion is brought about by gases such as CFCs. Holes in the ozone layer allow increased levels of radiation, causing increases in skin cancer and cataracts.
- Acid rain is brought about by the release mainly of sulfur dioxide through the burning of fossil fuels. Acid rain is damaging forests and killing organisms in rivers and lakes.
- Release of organic material, such as sewage, into rivers increases the biological oxygen demand, and reduces the number of species that can inhabit the river. The degree of pollution can be measured by the presence of certain indicator species.

- If inorganic fertilisers are overused, nitrate and phosphate ions may leach into waterways and cause nutrient enrichment — eutrophication. This increases the growth of algae; their subsequent death and decomposition by aerobic bacteria causes a fall in oxygen levels and many aquatic organisms die.
- Productivity in agriculture is increased by: monoculture; the use of fertilisers and pesticides; and intensive rearing of livestock. However, there are associated problems with these practices.
- Sustainable methods of agriculture are being promoted, for example polyculture and the use of organic fertilisers. Organic fertilisers maintain soil structure but are slow to act. Direction in the use of artificial (inorganic) fertilisers now occurs to reduce the risk of run-off and consequent eutrophication.
- Integrated pest management is recommended. This includes: use of resistant crop varieties; use of narrow-spectrum, non-persistent pesticides; and biological control — using a predator or parasite of the pest.

Questions & Answers

This section consists of two exemplar papers constructed in the same way as your A2 Unit 1 examination papers. Following each question, there are answers provided by two students of differing ability.

Examiner comments

Each question is followed by a brief analysis of what to look out for when answering the question (shown by the icon ⓔ). All student responses are then followed by examiner comments. These are preceded by the icon ⓔ. They provide the correct answers and indicate where difficulties for the student occurred. Difficulties might include lack of detail, lack of clarity, misconceptions, irrelevance, poor reading of questions and mistaken meanings of examination terms. The comments suggest areas for improvement.

Using this section

You could simply read this section, but it would be more effective to:
- try the questions before looking at students' responses and/or the examiner's comments, allowing yourself 2 hours for each paper
- check your answers against the students' responses and the examiner comments
- use the answers provided and the examiner comments to mark your paper
- use the students' responses and the examiner comments to check where your own performance might be improved.

Tips for answering questions

Make sure that your answer is sufficiently detailed. The examiners give you guidance about how much you need to write.
- Use the mark allocation. In general, the number of marks indicates the number of points that you should provide — for a question worth 4 marks you need to give more points than for one worth 2 marks.
- Use the recommended time allocation. You are advised to spend 25 minutes on Section B. Try to keep to this. It is possible to write a longer 'essay' but you may be providing more points than there are marks available.

Read the questions carefully. There are two aspects to this.
- Pay attention to the command terms used in each question — for example, 'describe' or 'explain' — and make sure you respond appropriately.
- The stem of a question may provide information needed to answer the question. Think about how this information can help you to construct or focus on a relevant answer.

Explanations of the command terms used in questions can be found in Appendix 1 of the specification and in the biology microsite at www.ccea.org.uk.

Exemplar paper 1

Section A

Question 1

(a) Read through the following passage about phytochromes, and write the most appropriate word(s)/numbers in the blank spaces to complete the account. (3 marks)

The photosensitive pigment phytochrome (P) exists in two interchangeable forms P_{660} and P_{730}. P_{660} absorbs _____ coloured light, while P_{730} absorbs _____ coloured light. During the day P_____ is converted to P_____ while, in the dark the process is reversed.

(b) The diagram below shows the flowering responses of a short-day plant and a long-day plant when exposed to a specific light regime.

| Light period | | | Dark period | | Flowering response |

Short-day plant does not flower

Long-day plant flowers

```
0        6        12        18        24
            Time/hours
```

Explain the different flowering responses stated for these two plant types in terms of P_{660} and P_{730}. (3 marks)

Total: 6 marks

ⓔ This is a straightforward question that you might find at the start of the paper. Part (a) tests recall (AO1) but, if you can remember that visible light varies in wavelength between 400 nm and 700 nm, then you can work it out. Part (b) requires more thought and the ability to express ideas well.

Student A

(a) red ✔, blue ✗, 660, 730 ✔ **a**
(b) The LDP flowers as there is enough light to convert the P_{660} to P_{730}, ✔ whereas the SDP doesn't flower as the period of time for darkness isn't long enough to convert P_{730} back to P_{660}. ✔ **b**

ⓔ **4/6 marks awarded a** While phytochrome is a blue pigment, it absorbs at the red end of the spectrum. P_{730} absorbs far-red light. 2 marks scored. **b** The student understands the interconversions of the two forms of phytochrome in this light regime and, while not precisely stating so, clearly understands that P_{730} stimulates flowering in LDPs. However, a clear statement that the accumulated P_{730} inhibits flowering in SDPs is lacking. It is important to understand that P_{730} is the active form, whether stimulatory or inhibitory. 2 marks scored.

Student B

(a) red ✔, far-red ✔, 660, 730 ✔ **a**

(b) The SDP does not flower because the dark period is less than the critical length and so P_{730}, which inhibit its flowering, ✔ is not sufficiently converted to P_{660}. ✔ A LDP, however, needs a short period of darkness, as here, to allow accumulation of P_{730} and so stimulate flowering. ✔ **b**

ⓔ **6/6 marks awarded a** Three correct answers. **b** The student has produced a well-worded answer showing understanding of the relevant levels of phytochrome for this light regime and of the different responses of long-day and short-day plants to relatively high levels of P_{730}. All 3 marks scored.

Question 2

The diagrams below show two images of a myofibril from a striated muscle fibre, as it would appear in transverse section.

(a) **Name the proteins which make up the filaments labelled 1 and 2.** (2 marks)

(b) **Explain why the images A and B are different from each other.** (2 marks)

(c) **Which of the images, A or B, would be more prominent in the myofibril of a contracted muscle fibre? Explain your reasoning.** (2 marks)

(d) **Explain the role of each of the following during the contraction of a muscle fibre.**
   (i) **Calcium** (1 mark)
   (ii) **ATP** (1 mark)

Total: 8 marks

ⓔ The first part of a question is often straightforward and this is no exception; you should not be troubled by part (a). Parts (b) and (c) require an evaluation of the diagrams; these are transverse sections of structures that may be more familiar to you as longitudinal sections. Part (d) is recall of the role of two chemicals during muscle contraction — you need to be selective in your answer and not just write all you know about contraction.

Student A

(a) 1. actin ✔ 2. myosin ✔ **a**
(b) Image A is a transverse section of the thick and thin filaments ✗ whereas image B is a transverse section of the I band. ✔ **b**
(c) Image A ✔ because when the muscle contracts the thick filaments become more prominent than the thin filaments. ✗ **c**
(d) (i) Calcium is involved in synaptic transmission stimulating muscle contraction. ✗ **d**
    (ii) ATP is needed to release energy to detach the myosin head from the actin. ✔ **e**

ⓔ **5/8 marks awarded a** Both answers are correct, for 2 marks. **b** The first point does not explain why both thick and thin filaments are apparent — it is a section through the A (dark) band though not the central H-zone component of it. The second point is correct for 1 mark. **c** This is not the correct reason. The region of overlapping filaments increases as the filaments slide over each other when the muscle contracts. 1 mark out of 2 scored. **d** This is not relevant: calcium is involved in initiating synaptic transmission from the synaptic bulbs; however, the question asks about its role during muscle contraction. No mark scored. **e** The answer is correct, for 1 mark.

Student B

(a) 1. actin ✔ 2. myosin ✔ **a**

(b) A is taken through the dark end of the A band, i.e. not through the H-zone where there are only thick filaments. ✔ B is taken through the I band where there are only thin filaments. ✔ **b**

(c) A, ✔ since when the muscle contracts the light bands and H-zones become shorter while the region of overlap increases as the actin filaments are drawn inwards over the myosin. ✔ **c**

(d) **(i)** Calcium ions from the sarcoplasmic reticulum initiate changes which free the actin for the attachment of myosin heads. ✔ **d**

    (ii) ATP attaches to the myosin head where ATPase releases the energy required for the head to swing in an arc, moving the actin filament. ✗ **e**

ⓔ **7/8 marks awarded a** Both answers are correct, for 2 marks. **b** This is a brief but correct answer which scores both marks. **c** This is a concise and complete answer which scores both marks. **d** This is correct, for 1 mark. **e** This is not correct. The rotation of the myosin head is caused by a conformational change in the protein as the head attaches to the actin. ATP is required for detachment. The answer fails to score a mark.

Question 3

(a) **The rabbit is a herbivore. The diagram below shows the daily energy budget for a rabbit. (Figures are kJ day⁻¹.)**

(i) **Write an equation for production (P) in the rabbit. Use the symbols shown in the diagram above.** (1 mark)

(ii) **Calculate the percentage of energy of the food consumed that is lost to decomposers (show your working).** (2 marks)

(iii) **The fox is a carnivore. The percentage of energy lost by a fox to decomposers is 15%. Explain why the rabbit and fox figures differ.** (1 mark)

(b) **Describe how the size of a rabbit population could be estimated using the capture–recapture technique.** (4 marks)

Total: 8 marks

ⓔ Part (a) requires you to study carefully the energy budget for a rabbit: in (a) (i) you can work out the equation for production — there are alternative correct answers; in (a) (ii) you must decide which components of the budget represent energy to decomposers; while, in (a) (iii) you need to compare figures for a herbivore and a carnivore — consider the type of food that they eat. Part (b) is AO3 and tests your understanding of the capture–recapture technique.

Student A

(a) (i) P = C − (R + U + F + A) ✗ **a**
(ii) 700 ✗ ÷ 1150 × 100% ✔ = 60.9% **b**
(iii) The faeces of the rabbit contain cellulose. ✔ **c**
(b) A number of rabbits are captured, marked and released back into the environment. A day later another sample of rabbits are captured ✔ and the number of marked ones counted. ✔ The number initially marked is multiplied by the number of recaptures to get the number in the population. ✗ **d**

ⓔ **4/8 marks awarded a** This is incorrect. A should have been ignored in the equation as written — it is a major part of consumption (C — i.e. that part of consumption not egested as faeces (F)) and includes production (P), respiration (R) and excretion (U). Production could be written to include assimilation (A) as P = A – R – U, which is an alternative answer to that supplied by Student B. No mark awarded. **b** The amount of material given as going to decomposers is incorrect since it ignores the urine. 1 mark is awarded for the correct denominator and procedure for calculating a percentage. Note that marks are awarded for procedures, not just for the correct answer. 1 mark out of 2 awarded. **c** There is sufficient here for 1 mark. The answer might have noted the difficulty of digesting cellulose. **d** The first sentence is not wrong but gives no suggestion as to a suitable method either for capturing or marking the rabbits. The last sentence gives no idea as to what should be multiplied or divided. Two marks are awarded for: noting that a subsequent sample is taken after a suitable period; and that the number marked in the subsequent sample, i.e. the number of recaptures, should be counted. The equation for determining population size is wrong. 2 marks out of 4 awarded.

Student B

(a) (i) P = C – F – U – R ✔ a
   (ii) (700 + 20) ✔ ÷ 1150 × 100% ✔ = 62.6% b
   (iii) There is relatively more faeces produced by the rabbit as it is a herbivore and its diet has a lot of cellulose which is not fully digested. ✔ c
(b) Some rabbits should be caught and marked with a permanent marker on their underside. ✔ This will mean that it isn't removed and does not make the rabbit more obvious to predators. They are then released back into their environment. The following day further rabbits are caught ✔ and the number that are recaptures, that is, marked, are counted. ✔ d

---

ⓔ **7/8 marks awarded** **a** The answer is correct, for 1 mark. **b** The answer is correct, for 2 marks. **c** The answer is correct, for 1 mark. **d** There are three valid points here. However, there is no suggestion as to how the population size would be estimated, e.g. by multiplying the first and second sample sizes and dividing by the number of recaptures. 3 marks out of 4 are awarded.

Question 4

The graph below shows the density of cones and rods across the retina of a mammalian eye.

(a) **Analyse the information in the graph, and use your own understanding to answer the questions that follow.**
    (i) Explain why there is no photoreception at point X on the graph. (1 mark)
    (ii) Explain why cones provide visual acuity at A, but not at B. (2 marks)
    (iii) Explain why rods cannot provide visual acuity, even at B. (1 mark)

(b) **Describe how rods provide light perception even when under conditions of low illumination.** (2 marks)

(c) **Describe how cones provide colour perception.** (2 marks)

Total: 8 marks

ⓔ Parts (b) and (c), testing understanding of rods and cones respectively, are more straightforward and you may wish to start with them. Part (a) requires you to apply your understanding — you must study the graph carefully and then use your understanding of rod and cone function to answer the questions asked.

Student A

(a) **(i)** There are no cones present at X — this is the blind spot. ✗ a
     (ii) Because there are more cones present at A than at B. ✔ b
     (iii) A bipolar neurone is stimulated by a number of rods. ✔ c
(b) Many rods connect onto one bipolar neurone and so have an additive effect. ✔ d
(c) There are three types of cone, some sensitive to red light, some to blue light and
    some to green light. ✔ e

ⓔ **4/8 marks awarded a** This is the blind spot but it contains *neither* cones *nor* rods. The answer fails to score. **b** This answer is worthy of 1 mark. A second mark required some understanding of the synapsing of each cone to an individual neurone, and the ability to discriminate points close together. 1 mark out of 2 awarded. **c** This answer is just sufficient to gain 1 mark. **d** This answer is correct for 1 mark. However, a complete answer would note either that a combination of rods together produces the required generator potential or that there is summation of the amount of transmitter substance released into the synapse with the neurone of the optic nerve. **e** This is correct but does not explain how other colours are perceived — that stimulation of combinations of cone types sends a pattern of impulses to the brain which perceives a particular colour. 1 mark out of 2 awarded.

Student B

(a) **(i)** This is the blind spot where there are no rods or cones. ✔ **a**

(ii) At A the cones are densely packed into the fovea. Each cone synapses to single neurones so that lots of impulses are sent to the brain — the image is not blurred and the brain can detect images stimulating cones which spatially are close together. ✔ At B, there are few cones and these lie further apart. ✔ **b**

(iii) Because they display retinal convergence. There are many but light falls on many rods synapsing with one neurone and so the images cannot be separated. ✔ **c**

(b) They have higher sensitivity since the pigment in rods is broken down at low light intensity. ✔ **d**

(c) There are three types of cone, each with a different iodopsin sensitive to blue, green or red. ✔ **e**

ⓔ **6/8 marks awarded** **a** This answer is correct, for 1 mark. **b** This is a complete answer, for 2 marks. **c** This is a full answer, for 1 mark. **d** This is an alternative answer to that of Student A, but, again, it lacks the detail that a generator potential would more readily be produced. 1 mark out of 2 awarded. **e** This answer is correct but, again, does not explain the perception of other colours such as purple or yellow. 1 mark out of 2 awarded.

Question 5

(a) (i) **Name the nitrogen-containing ion that is released as a result of the decomposition of dead organisms.** (1 mark)

   (ii) **Name the nitrogen-containing ion that is absorbed by plant roots.** (1 mark)

   (iii) **Name the group of soil bacteria that converts the nitrogen-containing ion released by decomposers to the nitrogen-containing ion absorbed by plants.** (1 mark)

(b) **The availability of an ion in the soil is directly related to its solubility, which in turn depends on the soil pH. The graph below shows the effect of pH on the solubility of four mineral ions (the thickness of the horizontal bands represents the solubility of each mineral ion). The three mineral ions most required by plants are those containing nitrogen (N), phosphorus (P) and potassium (K).**

   (i) **Using the information above, deduce the pH range that would be optimal for plant growth. Explain your answer.** (2 marks)

   (ii) **Using the information in the graph, explain why aluminium is readily leached out of the soil when rainfall is highly acidic.** (1 mark)

   (iii) **Outline the consequences of high concentrations of aluminium entering rivers and lakes.** (2 marks)

   (iv) **Outline what causes rainfall to be highly acidic.** (2 marks)

   Total: 10 marks

ⓔ Part (a) requires you to select the appropriate answer from your understanding of the nitrogen cycle. In part (b) you must study the graph of the solubility of four ions at different pHs. This is novel, so you must be confident in your ability to make deductions based on the information supplied to answer sub-parts (i) and (ii). Sub-parts (iii) and (iv) are more straightforward, but notice that you need to be succinct — you are asked only for an outline and the mark allocation is 2 in each case.

Student A

(a) (i) Ammonium ion. ✔ **a**
  (ii) Nitrite ion. ✗ **b**
  (iii) Nitrifying bacteria. ✔ **c**
(b) (i) pH 6.5 to 6.9 ✔ would be optimal as all three ions needed are at their most soluble. ✔ **d**
  (ii) Aluminium is highly soluble in acidic conditions. ✔ **e**
  (iii) Aluminium in rivers can cause fish to die. ✔ The ions are trapped in their gills and so they are unable to survive. ✗ **f**
  (iv) Nitrous and sulfurous gases in the air ✗ have been caused by combustion. These dissolve in rainwater and then fall as acid rain. ✔ **g**

ⓔ **7/10 marks awarded a** This is correct, for 1 mark. **b** Nitrate ion is the correct answer. No mark scored. **c** The answer is correct, for 1 mark. **d** The range is correct, as is the explanation — 2 marks. **e** The answer is correct, for 1 mark. **f** The first point is worth 1 mark. However, aluminium stimulates mucus production — it does not itself lead to asphyxiation. 1 mark out of 2 awarded. **g** It is oxides of nitrogen and sulfur that are released by combustion. The second point is worthy of a mark. 1 mark out of 2 awarded.

Student B

(a) (i) Ammonium ion. ✔ **a**
  (ii) Nitrate ion. ✔ **b**
  (iii) Nitrifying bacteria. ✔ **c**
(b) (i) pH 6.5 to 7 ✔ as this is when the N, P and K bands are thickest and so most soluble and more available for the plants. ✔ **d**
  (ii) Under acidic conditions the aluminium bar is very thick which means that aluminium is highly soluble in acid rainwater and will leach out of the soil. ✔ **e**
  (iii) High concentrations of aluminium cause fish to secrete mucus ✔ and this limits gaseous exchange so that the fish asphyxiate. ✔ **f**
  (iv) Nitrogen and sulfur dioxides are released as a result of the combustion of fossil fuels ✔ and these react with the moisture in the air to form nitrous and sulfurous acids ✔ which then fall as acid rain. **g**

ⓔ **10/10 marks awarded a–g** Correct and full answers throughout.

Question 6

With the aim of obtaining immunity from a serious viral disease affecting a group of patients, two nurses received injections. One nurse was given antiserum containing antibodies for the virus. The other nurse was given a vaccine.

The graph below shows the level of antibody in the blood plasma of the two nurses over a period of 36 weeks after the injections were given.

(a) **Which nurse was given a vaccination? Using the information in the graph and your own understanding, explain your answer.** (3 marks)

(b) **An antibody level of more than 60 arbitrary units is needed to provide immunity against the virus. For how many weeks does Nurse A remain immune?** (1 mark)

(c) **Nurse A was sent to treat patients showing the disease symptoms at the beginning of the first week. Nurse B was sent at a later date. Explain why Nurse B was able to remain treating patients for a longer period of time than Nurse A.** (3 marks)

Total: 7 marks

ⓔ Again this is novel information, so you must read the question stem and study the graph carefully before responding. Part (a) requires an explanation for the antibody levels in the two nurses — note that the mark allocation is 3. In part (b), you need to take care in reading 60 units on the *y*-axis to answer the question. To answer the question fully in part (c), you must explain the situation for *both* nurses.

(a) Nurse B, ✔ since some of the virus will have been included in the vaccine so that the nurse's body produced its own antibodies. ✔ a
(b) 9 weeks. ✗ b
(c) Because over time the antibodies in nurse A's blood would be destroyed as her own immune system would recognise them as foreign antigens. ✔ Also, since there were no plasma cells produced, no antibodies are being produced. ✔ c

ⓔ **4/7 marks awarded** a The answer is Nurse B, but this is not a full enough answer. 2 marks out of 3 are awarded. b The answer is incorrect and scores no marks. The scales are awkward, so more care is required. c These are well-made points, but the answer does not explain why antibodies are produced in Nurse B. 2 marks out of 3 are awarded.

(a) Nurse B. ✔ When the antigens are injected in the vaccine there are no antibodies since it takes time for the correct B lymphocytes to be activated ✔ and to produce plasma cells which in turn produce antibodies. ✔ a
(b) 8 weeks. ✔ b
(c) Nurse A has only received antibodies and so the immunity is passive. ✔ This is temporary immunity since no plasma or memory cells are produced ✔ and the antibodies are broken down over time. ✔ In nurse B, the immunity is long term since plasma cells produce antibodies over time ✔ and memory cells are produced which will respond to the antigen if it enters the body at a later stage. ✔ c

ⓔ **7/7 marks awarded** a This is a full answer, for a score of 3 marks. b The answer is correct, for 1 mark c A very full answer, worth the full 3 marks.

Question 7

Daphnia, commonly called water-fleas, are small, planktonic crustaceans found in fresh water, where they mainly feed on algae. *Daphnia* may, in turn, be preyed upon by *Hydra*.

(a) The graph shows the numbers of *Daphnia* and *Hydra* in a pond over a 9-week period. Explain the changes in both the prey (*Daphnia*) and predator (*Hydra*) populations.

(4 marks)

(b) Populations of *Daphnia* are readily grown in laboratory tanks containing an ample source of algae (e.g. *Chlamydomonas*). The graph below shows the growth of two populations of *Daphnia* at two different temperatures.

(i)   Continue the curves on the graph to show what you would expect to happen to each
      population in the next 20 days.   (2 marks)
(ii) Explain the curve that you have drawn for the population at 18°C.   (2 marks)

(c) *Daphnia* have been used to test the effects of toxins on an ecosystem. The chemical cadmium is
   known to have a negative effect on the growth of *Daphnia* populations, though it is suggested
   that high food levels protect *Daphnia* from the adverse effects of the toxin. Design
   an experiment to test this hypothesis.   (4 marks)

Total: 12 marks

ⓔ This question is about different aspects of population growth. In part (a) you need to explain
predator–prey curves — notice that the mark allocation of 4 indicates that a full answer is
required. In part (b) you need to work out the effect of temperature on population growth. Does
temperature influence the rate of growth, or the carrying capacity, or both? In part (c) it does not
matter if you have heard of *Daphnia* or cadmium. You need to apply the principles of experimental
design, for example: what factor you would vary (the independent variable); the need for a control
experiment; what variables you would control (not the same thing as a control experiment); and
what you would measure (the dependent variable).

Student A

(a) Prey: In the first few weeks the *Daphnia* reproduce and multiply rapidly. ✗ It peaks
    at 3 weeks and then numbers rapidly fall, as they are now being predated by large
    numbers of *Hydra*. ✔
    Predator: The *Hydra* population doesn't start to increase until there are high
    numbers of prey and there is plenty of food. ✔ Their numbers then drop as the
    food source is depleted. ✔ **a**
(b) (i) The graph shows a curve for 23°C that levels off at a density of approximately
        290 ✔; and a curve for 18°C that levels off at around 170. ✗ **b**
    (ii) It takes time for the *Daphnia* population to get used to the habitat ✗ and for the
         algae to start growing. There would be a lack of food for the *Daphnia*. ✗ **c**
(c) Two separate tanks are set up, each containing a certain amount of cadmium. ✔
    Daphnia are added to both. One of the tanks is left with a limited food source ✔
    and the results recorded, while the other has a vast food source and the results
    recorded. ✔ **d**

ⓔ **7/12 marks awarded a** Prey: the first sentence lacks any reference to lack of environmental
resistance, in this case the absence of predators. The second point is correct. Predator: both points
are worthy of marks. 3 marks out of 4 scored. **b** The prediction at 23°C is correct. The prediction
for 18°C is incorrect — temperature will not limit the carrying capacity, only the rate of increase.
1 mark out of 2 scored. **c** This answer does not seem relevant and scores no marks. The lower
temperature will have a direct effect on the rate of development of the *Daphnia* population but
will not influence food supply. **d** There is enough here for 3 marks out of 4, but there are many
other possible points: growing an inoculum population of *Daphnia* so that the number of *Daphnia*
added to each tank is comparable; growing algae to obtain a high-density population; a control
set-up with no cadmium added; other variables kept the same; statistical tests to compare final
densities.

Student B

(a) Prey: Due to the initial absence of the predator, *Hydra*, the prey can reproduce quickly. ✔ However, when the *Hydra* population increases the prey population falls as the *Hydra* eats them, thus raising the death rate. ✔
Predator: When the amount of prey is sufficient, *Hydra* arrive in the area, feed on the prey, reproduce and increase in number. The lag is due to the time for reproduction. ✔ When prey numbers start to decrease, the predators have little to eat and so their population declines. ✔ **a**

(b) (i) The graph shows a curve for 23°C that stabilises around a density of approximately 290 ✔; while the curve for 18°C increases gradually towards 290. ✔ **b**

(ii) At 18°C, the metabolism of the *Daphnia* is slower and so they do not develop as quickly. ✔ However, they still have 'an ample source of algae' so they reach the same carrying capacity, but more slowly than at 23°C. ✔ **c**

(c) Place the *Daphnia* in a tank with an ample supply of algae and add the chemical cadmium. ✔ At the same time, place the same number of *Daphnia* ✔ in a tank with no algae, i.e. no food supply. ✗ After a period of time, count the number of *Daphnia* in each tank and record results. ✗ **d**

ⓔ **10/12 marks awarded a** This answer shows a thorough understanding of predator–prey oscillations and scores 4 marks. **b** Both predictions are correct, for 2 marks. **c** This is a correct and well-explained answer, for 2 marks. **d** There are only two relevant points here. The *Daphnia* would not survive if there was no food, never mind the cadmium. The reference to recording results is too vague. Other relevant points are noted in the examiner response to Student A.

Question 8

(a) The figure below is a photomicrograph of a section through part of the kidney cortex. The magnification of the micrograph is ×200.

MANFRED KAGE/SPL

Identify the structures labelled **A** to **C** in the photograph. *(3 marks)*

(b) Explain the process of ultrafiltration in the kidney cortex. *(3 marks)*

(c) Explain how water is reabsorbed from the filtrate. *(3 marks)*

(d) The graph below shows the amount of water and urea in filtrate and in urine (mean daily values are given).

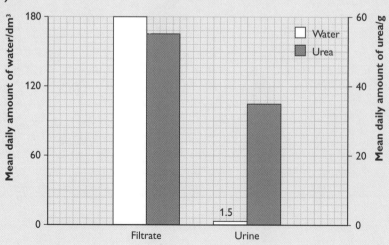

(i) **Use the information in the graph to calculate how many more times urea is concentrated in urine than in the filtrate (show your working).** (3 marks)

(ii) **Explain the importance of the reabsorption of water.** (1 mark)

Total: 13 marks

ⓔ You may not have seen a photograph like this before but, significantly, you are told that it is from the kidney cortex and that the magnification is ×200, so you should be able, in part (a), to work out what features A to C represent. If you cannot answer part (a) immediately, then leave it and come back to it later after answering other parts on kidney function. Parts (b) and (c) require more straightforward knowledge of aspects of the functioning of the kidney: ultrafiltration in (b); and water reabsorption in (c). Part (d) involves obtaining values from the graph to undertake a calculation. There are alternative ways of tackling this and you should be aware that the calculation is more a matter of common sense than of mathematical ability.

Student A

(a) A: nephron ✗, B: cortex ✗, C: mitochondrion. ✗ **a**

(b) Ultrafiltration is the forcing of small molecules in the plasma, but not the protein or blood cells, through the wall of the glomerulus. ✔ The actual filter is the basement membrane ✔, since both the glomerulus and the capsule wall are leaky. **b**

(c) 80% of the water is actually reabsorbed in the proximal convoluted tubule. ✔ More water is reabsorbed from the collecting duct so that the urine is hypertonic. ✔ This happens as a result of the loop of Henlé making the lower reaches of the medulla more salty. ✔ **c**

(d) (i) In filtrate, the concentration of urea is 55 ÷ 180 = 0.306 ✔
In urine, the concentration of urea is 35 ÷ 1.5 = 23.333 ✔
Overall factor of concentration is 23.333 ÷ 0.306 = 76.3 ✔ **d**

(ii) If it wasn't reabsorbed sufficiently, then too much water would leave the body which would become dehydrated. ✔ **e**

ⓔ **9/13 marks awarded** **a** There would appear to have been no regard to the information in the stem: the section is of the cortex region of a kidney, while a mitochondrion would be simply too small to appear in the section. The answer scores no marks. **b** There are two correct points here. However, there is no reference to the build-up of pressure in the glomerulus. 2 marks out of 3 scored. **c** This answer is sufficient for 3 marks. **d** The answer is correct, and well laid out, for 3 marks. **e** The answer is correct, for 1 mark.

Student B

(a) A: cortex ✗, B: medulla ✗. C: blood capillary. ✗ **a**

(b) Blood enters the glomerulus via the afferent arteriole, which is wider in diameter than the efferent arteriole, creating a hydrostatic pressure. ✔ This pressure causes small molecules to be filtered from the glomerulus into the Bowman's capsule, ✔ while large molecules like proteins are too large to pass through. **b**

(c) In the limbs of the loop of Henlé, movement of ions and water result in the lower regions of the medulla containing more salt and so having a lower water potential. ✔ Thus an osmotic gradient is established through the medulla so that water is drawn out as the filtrate passes down the collecting duct. ✔ c

(d) **(i)** 55 g in 180 dm^{-3} and 35 g in 1.5 dm^{-3} ✔

equates to 2100 ÷ 55 ✗ = 38 ✗ d

(ii) To produce a hypertonic urine and conserve water in the body. ✔ e

ⓔ **6/13 marks awarded** **a** The answers are 'glomerulus', 'Bowman's capsule' and 'convoluted tubule' respectively. Recognition of structures is often poorly done. The student needs to make full use of the information in the stem, including the magnification. The answer scores no marks. **b** There is no reference to the nature of the filter — effectively the basement membrane. 2 marks out of 3 scored. **c** Two full points are presented, but the student should be aware than most water reabsorption occurs by osmosis in the proximal convoluted tubule. 2 marks out of 3 awarded. **d** The values for the amount of urea in volumes of filtrate and urine have been read from the graph correctly. However, it is impossible to determine where the value of 2100 came from and the answer supplied is incorrect. 1 mark only awarded for reading values off the graph. **e** The answer is correct, for 1 mark.

ⓔ **Section A total for Student A: 44 marks out of 72. Section A total for Student B: 59 marks out of 72.**

Section B

Question 9

> **The demand by the public for cheap, readily available food led to intensive farming practices. Over the past 60 years, there have been great increases in farm yields mostly as a result of the use of pesticides and the use of artificial fertilisers. However, there is an environmental cost to these practices, and there have been noted biodiversity issues and pollution effects.**
>
> **Discuss the 'biodiversity issues and pollution effects', and the remedies provided by sustainable farming, with respect to the use of**
> - **pesticides**
> - **artificial fertiliser** (16 marks)
>
> **Quality of written communication is awarded a maximum of 2 marks in this section.** (2 marks)
>
> **Total: 18 marks**

ⓔ In this question, worth 18 marks, you are required to write in continuous prose because 2 marks are awarded for quality of written communication (QWC). This is essentially an essay and it is best that you organise your thoughts before answering — you can do this in an *essay plan*. Notice that you must cover *two* issues — biodiversity *and* pollution — influenced by each of *two* features of intensive farming — the use of pesticides *and* artificial fertilisers. The mark scheme will insist you cover each of these areas, even if this is not detailed in the question. You must be able to sequence your ideas and emphasise the links between cause and effect, which is how QWC will be assessed.

Student A

A pesticide is a man-made chemical deliberately introduced into an environment to kill pests. Pesticides include herbicides, fungicides and insecticides. The first application of pesticides may be successful, but more problems may occur after the first application. Target pest resurgence occurs when the pesticide kills natural predators so that the pest may grow to even greater numbers. ✔ There can also be secondary pest outbreaks where a previously uncommon species increases in number because its competitor, the pest species, has been removed. ✔ In addition, a mutation for resistance may occur so that resistant forms become common in the pest population. ✔ Farmers then have to use higher concentrations of pesticides and apply them frequently, a condition known as pesticide addiction. ✔ This can be dangerous if pesticides get into the food chain. A pesticide may be stored in the fatty tissue of animals ✔ and, as a result of animals higher in the food chain eating more of those at the lower levels, the pesticide accumulates in top carnivores to toxic concentrations. ✔

If farmers depend on the use of artificial fertilisers it can result in the loss of crumb structure ✔ as well as an increased risk of eutrophication. ✔ These fertilisers leach into nearby rivers where they reduce the biological oxygen demand, killing fish and invertebrate life. ✗ a

QWC **b**

ⓔ **9/18 marks awarded** a Eight appropriate points are provided. However, there is a lack of balance in the essay. Information with respect to problems associated with the use of artificial fertilisers is limited. Indeed, in the last sentence there is obvious confusion between fertiliser run-off and organic pollution as causes of eutrophication. Of even greater significance, there is no attempt to describe remedies that might be provided through the practice of sustainable farming, e.g. biological control, use of narrow-spectrum non-persistent pesticides, use of manure and restrictions in the use of artificial fertilisers. b While ideas are clearly expressed, and the account is reasonably well sequenced, the lack of balance means that there are insufficient links between cause and effect, so the student scores only 1 mark for the quality of written communication.

Student B

Pesticides are commonplace on most farms as a way of controlling pests. However, problems arising from using them include: target pest resurgence, caused by removal of natural predators of the pest; ✔ secondary pest outbreaks, whereby a previously uncommon species booms with the removal of its natural enemies; ✔ and pesticide resistance. This leads to pesticide addiction, as greater and more frequent amounts of pesticide have to be applied. ✔ Pesticides have an adverse effect on the biodiversity of an area especially with the use of indiscriminating broad-spectrum pesticides. ✔ Furthermore, some pesticides, such as DDT, are persistent and are stored in animal fat, ✔ passing through the trophic levels to reach lethal concentrations. ✔

Artificial fertilisers also pose a threat to the environment. Minerals, such as aluminium, can become deposited on the gills of fish. ✗ The fish produce mucus against this, which clogs the gills causing asphyxiation. ✗ They can also cause nutrient enrichment — eutrophication — of waterways and algal blooms. When the algae die they represent a large biological oxygen demand for decomposing bacteria ✔ and the resultant anoxic conditions lead to fish and invertebrate death. ✔ Also a health risk is introduced by the possibility of drinking water being contaminated by certain toxic algae. ✔

In sustainable agriculture, a biological control agent is an alternative to a pesticide. Although slower in taking effect biological control can be long-term. When a more rapid response is required, then narrow-spectrum pesticides, which target only the pest, can be used. ✔ As an alternative to artificial fertiliser, organic fertiliser such as farmyard manure is beneficial since it improves soil aeration and drainage. ✔ Also by using a wide variety of crop plants and rotating them annually it helps inhibit pests from becoming established. ✔ Crop rotation also ensures that the soil is not depleted of any particular mineral, while a crop of legumes will improve the nitrogen content of the soil. ✔ a

QWC b

ⓔ **15/18 marks awarded** a While there has been some confusion between the effects of acid rain and eutrophication, this is a reasonably comprehensive and balanced essay. Some references, for example to 'pest resistance' (line 5) and to 'biological control agent' (line 18), are not sufficiently explained to gain a mark. There are 13 valid points. b This is, generally, a well-structured account and ideas are expressed fluently, so 2 marks are awarded for the quality of written communication.

(e) Section B total for Student A: 9 marks out of 18. Section B total for Student B: 15 marks out of 18

Paper total for Student A: 53 marks out of 90, which equates to a grade C. While Student A has demonstrated a reasonable knowledge from across the course, there is a lack of detailed understanding throughout. This would indicate that while revision of the course was undertaken, it could have been more thorough. The student has also exhibited the skills of analysis and evaluation, and has developed the confidence to deal with unfamiliar situations such as those in questions 5 to 8. Numeracy skills are good, though marred by what may be presumed to be 'slips' (question 3 (a) (ii) and question 6 (b)). It is always advisable to leave some time to check arithmetic work and especially when reading values off a graph. A few extra marks may be sufficient to raise the grade level. Overall, this is a mid-grade performance which could have been improved by developing 'active learning' strategies, such as those indicated in the introductory section to this book, for use during the revision process.

Paper total for Student B: 74 marks out of 90, which is a grade A performance. Student B has shown a thorough understanding of the course. In particular, unfamiliar contexts, the basis for many A2 questions, were readily dealt with. (S)he uses the information — written, tabular or graphical — in question stems to provide detailed answers. The skills of analysis and evaluation are well developed and, while (s)he found difficulty with question 8 (d) (i), it is not a sufficient weakness to indicate that numeracy skills are lacking or that this is not a top-level performance.

Exemplar paper 2

Section A

Question 1

(a) Describe what is meant by the term 'pioneer species'. (1 mark)

(b) Define the term 'climatic climax'. (2 marks)

(c) Compare and contrast primary and secondary succession. (2 marks)

Total: 5 marks

ⓔ This is straightforward, as expected at the start of a paper. All parts require understanding of terms used in the succession topic. You will be advantaged if you have prepared a glossary of terms for the topic. Part (c) uses the command *compare and contrast* — that means that you must identify a *similarity* and a *difference*.

Student A

(a) The first species to colonise bare ground. ✔ a
(b) Climatic climax is when the species composition is determined by climatic factors, such as rainfall. ✔ b
(c) Both successions involve community development. ✗ Primary succession (slow) occurs in a previously unoccupied area, e.g. bare rock, while secondary succession (faster) occurs when the community has been destroyed, e.g. by fire, but where there is still a soil structure. ✔ c

ⓔ **3/5 marks awarded a** Correct for 1 mark. **b** This is fine for 1 mark, but there is no explanation of 'climax', i.e. that it is the final, relatively stable community or that it is at the end-point in succession. **c** The similarity is just too vague to be awarded a mark. The difference is correct. 1 mark out of 2 scored.

Student B

(a) The initial species that occupies an area of previously unoccupied land. ✔ a
(b) This is when the composition of a stable community, at the end of succession, ✔ is determined by climatic factors such as temperature and water availability. ✔ b
(c) Primary succession is a relatively slow process of community development, while secondary succession is a much more rapid process. ✗ Both involve changes in the number and types of species in the developing community over time. ✔ c

(e) **4/5 marks awarded** **a** Correct, for 1 mark. **b** This is correct with respect to both 'climatic' and 'climax', for 2 marks. **c** The contrast does not make reference to the crucial point about whether or not there was a community occupying the area beforehand. The comparison is correct. 1 mark out of 2 scored.

Question 2

Viruses may invade the body and infect certain host cells, in which they proliferate and are then released to infect further host cells. The diagram below represents a virus-infected host cell and free viruses that have been released from an infected cell.

Viral antigen

Free viruses

Virus-infected host cell

(a) Explain how cell-mediated immunity within the body leads to the destruction of virus-infected cells. **(4 marks)**

(b) Explain how antibody-mediated immunity leads to the destruction of free viruses. **(4 marks)**

Total: 8 marks

ⓔ If you understand the topic, you shouldn't really need the diagram to answer either part (a) or (b). However, the diagram emphasises why the two forms of immunity to antigens are necessary and draws the topic together. Notice that the mark allocation for each question part is 4 and so you will need to offer full explanations — from detection to destruction.

Student A

(a) Cell-mediated immunity involves T lymphocytes. The T-cell has a specific receptor which attracts the antigen. ✗ The T-cell is activated and divides by mitosis to produce a clone. ✔ The cells of the clone differentiate to form 4 cell types: T-suppressor; T-killer; T-helper; and T-memory. ✔ The T-killer cells engulf the infected cells. ✗ **a**

(b) Antibody-mediated immunity is carried out by B lymphocytes. The B-lymphocyte is attracted by a specific viral antigen. ✗ The B-cell responds by dividing to produce a clone. ✔ Among the cells of the clone plasma cells and memory cells are formed. ✔ Plasma cells produce the antibodies ✔ which then destroy the free viruses. **b**

ⓔ **5/8 marks awarded a** There are two correct points here. There is no 'attraction' between the receptor and the antigen — they have a complementary shape; and, specifically, one of the many T-cells available has the complementary receptor. Further, killer T-cells do not destroy infected cells by phagocytosis. **b** There are three correct points here.

CCEA A2 Biology

Student B

(a) This involves the activity of T lymphocytes. The binding of the viral antigen and the complementary receptor of a particular T-cell ✔ initiates mitotic division of the T-cell. ✔ The cells of the resultant clone develop into one of four types of cell, including helper T-cells, killer T-cells, memory T-cells and suppressor T-cells. ✔ Killer T-cells produce perforins which produce perforations in the surface membrane of the infected host cell. ✔ Helper T-cells activate plasma cells and killer T-cells, and stimulate the activity of phagocytes such as macrophages. ✔ **a**

(b) The binding of an antigen on the surface of a free virus and the complementary receptor site of a specific B-cell ✔ causes an antibody-mediated response. The B-cell enlarges and divides mitotically to produce a clone. ✔ The cells of the clone develop into either memory B-cells or plasma cells. ✔ Plasma cells synthesise and secrete antibodies which are specific and bind to the antigens of the viruses. ✔ The response may involve agglutination or opsonisation ✔ followed by phagocytosis. **b**

ⓔ **8/8 marks awarded a** Four (and more) correct points. **b** Four correct points are included in a good, well-worded account.

Question 3

The robin (*Erithacus rubecula*) is a familiar bird — its red breast plumage, song and aggressiveness to other robins are important in asserting territorial rights.

Ringing experiments of a population of robins over many years have allowed the study of the mortality of different age classes. The study involved 130 robins and the results are shown in the table below.

Age/years	Number surviving at beginning of age interval	Number dying in age interval
0–1	130	94
1–2	36	17
2–3	X	14
3–4	5	Y
4–5	2	0
5–6	2	0
6–7	2	0
7–8	2	0
8–9	1	1

(a) Determine the missing figures, **X** and **Y**, in the table. (2 marks)

(b) A pair of robins have an average clutch size of five eggs, and two clutches are produced per year. Using the information in the table, calculate the number of young birds, produced by a pair of robins, which might be expected to survive their first year (show your working). (2 marks)

(c) It has been claimed that natural selection has favoured early reproductive maturity in robins. Suggest an explanation for this claim. (2 marks)

(d) Early maturation is a feature of r-selected (opportunistic) species. Note one other characteristic of r-selected species. (1 mark)

Total: 7 marks

ⓔ You are unlikely to have been presented with this information before, but you must expect situations that are unfamiliar — they are testing your AO2 skills. You will need to study the table of data carefully and gain an understanding of the survival of robins as they get older. Only then will you be able to answer parts (a) and (b). In part (b), you are asked to *show your working* because I mark may be available for a correct arithmetic operation even if the final answer is incorrect. Part (c) requires an overall appraisal of the information, while part (d) is simple recall (AOI).

Student A

(a) X = 36 − 17 = 19 ✔; Y = 5 − 3 = 2 ✗ a

(b) Survival value in the first year is 36 ÷ 130 = 0.277 ✔
Two clutches of five = 10 young, which means 10 × 0.277 survive = 2.8 birds survive ✔ b

(c) Early maturation is needed to ensure a maximal reproductive rate. ✔ c

(d) They are easily adaptable. ✗ d

@ **4/7 marks awarded** a X is correct, but Y is not. This looks like a slip and shows the value of always checking your calculations. 1 mark out of 2 scored. **b** This is a correct expectation, for a score of 2 marks. **c** This is a correct point, but the student should have added that this was 'to counteract the very high rates of mortality in the early years'. The answer scores 1 mark out of 2. **d** This is too vague. There is a long list of r-selected features. No mark awarded.

Student B

(a) X = 19 ✔; Y = 3 ✔ a
(b) Mortality in the first year is 94 ÷ 130 × 100% = 72.3% ✔
72.3% of 10 young robins = 7.23, meaning 8 individuals die and so 2 survive ✔ b
(c) A high percentage of robins die early in their life ✔ so if they are to reproduce, and pass on their genes, before they die, ✔ early maturation is essential. c
(d) r-selected species produce large numbers of offspring. ✔ d

@ **7/7 marks awarded** a Both answers are correct, for 2 marks. **b** Student B has solved the problem differently from Student A, and even though a different answer has been presented the reasoning has been correct, scoring the full 2 marks. **c** This is correct, for 2 marks. **d** This is correct, for 1 mark.

Question 4

An oat seedling has a protective sheath (the coleoptile) which is frequently used as convenient plant material for experiments on auxin production and phototropism.

(a) An experiment was carried out in which the tips of oat coleoptiles were removed and placed on blocks of agar. These were either kept in the dark or left in the light.
The results of this experiment, with the relative amounts of auxin collected in the agar blocks, are shown in the diagrams below.

What can you deduce from this experiment about the production of auxin? (2 marks)

(b) In a second experiment, coleoptile tips were similarly placed on agar blocks. In this set-up, however, the agar blocks were divided and separated by thin sheets of metal and were kept either in the dark or illuminated from the right side.
The results of this experiment are shown in the diagram below.

What can you deduce from this experiment about the movement of auxin through coleoptile sections? Comment on the effect of light. (3 marks)

(c) In a third experiment an agar block containing auxin was placed asymmetrically on a decapitated coleoptile as shown in the diagram below.

Describe the result that you would expect, and explain how this would be brought about. (3 marks)

Total: 8 marks

ⓔ In this question, you are required to analyse three experiments involving auxins. In each case, you must read the information carefully and apply your understanding. In parts (a) and (b), be careful to base your answers on the information provided. In part (c), you must describe — put into words — the expected result and then offer an explanation, i.e. use your understanding to reason out what will happen.

Student A

(a) Auxin is produced in the tip ✔ and the amount produced does not depend on whether it is light or dark. ✔ **a**

(b) In the dark auxin does not move. ✗ However, light causes lateral movement of auxin. ✔ **b**

(c) The coleoptile would bend towards the side where there was no agar. ✔ This is due to auxin moving down the agar side ✔ and causing greater cell elongation and growth on that side. ✔ **c**

ⓔ **6/8 marks awarded a** The answer is correct, for a score of 2 marks. **b** This answer is too brief and the point in the first sentence is not strictly correct, since the auxin moves down from the tip. 1 mark out of 3 awarded. **c** There are three correct points here, even if not that well expressed. 3 marks awarded.

Student B

(a) Light intensity does not affect auxin production. ✔ **a**

(b) Auxin produced at the tip moves downwards. ✔ In conditions of unidirectional light, there is a lateral displacement of auxin ✔ to the non-illuminated side. ✔ **b**

(c) The coleoptile would grow up and to the right. ✔ This is because auxin will move down the coleoptile from the agar so that it becomes more concentrated on the left side. ✔ Auxin causes the elasticity of the cell wall to increase so that the cells on the left elongate more and grow more than on the right side. ✔ **c**

ⓔ **7/8 marks awarded a** Only one correct point has been given. 1 mark out of 2 awarded. **b** This is a full and correct answer, for 3 marks. **c** This is more detailed than the answer provided by Student A, for a score of 3 marks.

Question 5

The diagram below represents a vertical section through a mammalian eye.

Lens

Iris

Suspensory
ligament

Ciliary
muscle

(a) Describe the role of the iris when exposed to bright light and explain how it operates. (3 marks)

(b) Explain how the ciliary body, suspensory ligaments and lens function during accommodation
of the eye when viewing a distant object. (3 marks)

(c) Occasionally proteins in the lens clump together forming cataracts, which impair vision.
This is most commonly a result of aging but appears to be influenced by a number
of factors including exposure to excessive **UV light**. Explain how the amount of **UV light**
reaching the Earth's surface may be increasing as a result of environmental pollution. (3 marks)

Total: 9 marks

ⓔ The diagram is provided as stimulus material, but, if you know how the parts of the eye
function, you could probably answer the question without it. Part (a) is about the iris, describing
its role, i.e. what its purpose is, and then explaining how it operates; notice that the answer must
relate to the stimulus of *bright light*. In part (b), you must explain how accommodation of the lens
operates, taking care to ensure that this is in relation to a *distant object*. Part (c) is in the context of
the eye, but is really about one form of environmental pollution. Students frequently confuse the
different forms of atmospheric pollution so your answer needs to be specifically about UV light.

Student A

(a) In bright light the pupil becomes smaller ✔ so that less light enters the eye and protects the retina from damage. ✔ **a**

(b) The light rays from a far object are roughly parallel. ✔ The ligaments contract ✗ making the lens thinner and less refracting. ✔ **b**

(c) UV light reaching the Earth's surface is increasing because of ozone depletion ✔ due the release of CFCs and other gases such as oxides of nitrogen. ✔ As a result there is a hole in the ozone layer over Antarctica, which is having a particularly harmful effect on the people of Australia. **c**

ⓔ **6/9 marks awarded a** There are two correct points, but the student has not explained the operation of the iris, so scores 2 marks out of 3. **b** There are two correct points. Note that ligaments don't contract — they consist of connective tissue, which simply transfers any tension in the wall of the eyeball through to the lens. **c** Two correct points are made but the answer lacks any reference to the source of CFCs.

Student B

(a) In bright light, the iris restricts the entry of light so preventing damage to the rods and cones. ✔ It does this by contraction of the circular muscles and relaxation of radial muscles ✔ which makes the iris bigger but constricts the pupil. ✔ **a**

(b) When the object is distant, light rays are parallel ✔ and so don't have to be bent as much to be focused onto the retina. The muscles in the ciliary body relax ✔ tightening the suspensory ligaments ✗ and flattening the lens. **b**

(c) The amount of UV radiation reaching the Earth's surface is increasing as a result of a thinning of the ozone layer in the stratosphere. ✔ The ozone layer absorbs much of the UV radiation. This thinning is due to gases, such as CFCs, which cause ozone to break down. ✔ CFCs were used as coolants in refrigerators and as propellants in aerosols. ✔ While now no longer produced, there are large quantities of CFCs in existent refrigeration units. Furthermore, CFCs molecules are especially stable and may persist for up to 100 years. **c**

ⓔ **8/9 marks awarded a** The answer is correct and well expressed, for 3 marks. **b** The relaxation of the muscles in the ciliary body does not directly tighten the suspensory ligaments; rather their relaxation allows the tension in the wall of the eyeball (it is full of fluid) to be transferred through the suspensory ligaments. Two correct points have been made, for a score of 2 marks out of 3. **c** A complete answer, for 3 marks.

Question 6

(a) Carbon dioxide is a product of respiration. However, it is also produced by the combustion of fossil fuels and so atmospheric carbon dioxide levels have increased. It is known as a 'greenhouse gas'. Name one other greenhouse gas, and state a major source of the gas. **(2 marks)**

(b) A local council commissioned a survey to find out how much carbon dioxide was being produced in the council area. The table below shows the amount of carbon emitted by different sources over a one-year period.

Source	CO_2 production/10^3 tonnes y^{-1}
People and pets	8
Domestic fuels	22
Vehicles	30
Industrial fuels	40

   (i) Calculate the proportion of the total carbon dioxide emissions that is produced by vehicles (show your working). **(1 mark)**

   (ii) Describe one way in which vehicular emission of carbon dioxide might be reduced, and suggest why it would be difficult to predict that this would happen. **(2 marks)**

The council wished to initiate a project of planting fast-growing willow trees to absorb carbon dioxide. It has been determined that, after 10 years of growth, one hectare of planted willows will absorb 50×10^3 tonnes of carbon dioxide.

   (iii) Calculate the area of willow trees that would need to be planted to absorb the council area's carbon dioxide emissions over a 10-year period. (Show your working.) **(2 marks)**

   (iv) Explain the purpose of the council's willow project. **(2 marks)**

(c) Discuss the possible effects of global warming on the distribution and abundance of wild animals and plants. **(3 marks)**

Total: 12 marks

ⓔ The initial part (a), about greenhouse gases other than CO_2, is simple recall. In part (b) you need to use the information, on different sources of CO_2 emission in a locality and also on absorption of CO_2 by planted willow, to make a series of calculations and evaluations. Part (c) is recall of the effects of global warming but you must ensure that you keep your answer relevant by discussing only the effects on the 'distribution and abundance' of organisms. Parts (a) and (c) involve straightforward knowledge and understanding (AO1) and you might find it easier to answer these *before* answering part (b), which is AO2/AO3. Analysing questions in this way should be part of an examination strategy that you are developing.

Student A

(a) Methane. ✔ Breakdown, by fermentation, of plant material. ✔ **a**

(b) **(i)** 40 + 30 + 22 + 8 = 100, so 30 ÷ 100 = 0.3, and so 0.3×10^3 tonnes year⁻¹ ✗ **b**

   (ii) Using a renewable energy source. ✗ **c**

   (iii) Council emissions are 1000×10^3 tonnes of CO_2. ✗

      One hectare of willow absorbs 50×10^3 tonnes of CO_2, so dividing total emissions by this figure ✔ gives 2 hectares. **d**

(iv) The council's project would reduce the amounts of CO_2 in the atmosphere ✔ and so reduce global warming. **e**

(c) Global warming may cause species adapted to cooler conditions to become extinct in temperate regions. ✔ Further, species adapted to very cold conditions may suffer depleted numbers and a contraction of their distribution. ✔ The numbers of polar bears and the range over which they live will be restricted as the Arctic ice melts. ✔ Species adapted to warmer conditions may become more common in temperate regions. ✔ For example, several species of butterfly, previously restricted to the south of England, are now found in northerly regions of Britain. ✔ **f**

ⓔ **7/12 marks awarded** **a** Both points are correct. **b** This is incorrect, since the question asks for the proportion produced by vehicles, not the amount. The answer fails to score. **c** There is no explanation of how this energy might be made available for vehicular use. The answer scores no marks. **d** The willow figure represents 10 years' absorption so the emissions needed to be multiplied by 10. There is only one mistake, otherwise the procedure is correct, so 1 mark out of 2 scored. **e** The answer scores 1 mark out of 2. The student would really need to say something about this compensating for the extra CO_2 produced from, for example, vehicular emissions. **f** This is a full answer, and obviously a topic of particular interest to the student, scoring the full 3 marks.

Student B

(a) Sulfur dioxide ✗ produced by the burning of fossil fuels. ✗ **a**
(b) **(i)** $30 \div 100 = 0.3$ ✔ **b**
   (ii) Improve public transport so more people would use it, so that there would be fewer cars on the road and hence less vehicular emission of carbon dioxide. ✔ It would be difficult to predict that this would happen as you cannot force people to use public transport. ✔ **c**
   (iii) 50×10^3 tonnes of CO_2 are absorbed over 10 years, and the council's emission of CO_2 are 1000×10^3 tonnes over 10 years, ✔ so $1000 \div 50 = 20$ hectares of willow are required. ✔ **d**
   (iv) The project aims to use the willow trees to absorb CO_2 via their photosynthesis ✔ and in particular to remove that volume of CO_2 produced by vehicles. ✔ **e**
(c) A generally warmer climate may also influence when certain birds migrate from one region to another. ✔ It may also result in a return of the malarial mosquito to Europe. ✔ **f**

ⓔ **9/12 marks awarded** **a** Sulfur dioxide is associated with acid rain, but it is not a noted greenhouse gas. The answer scores no marks. **b** The answer is correct, for 1 mark. **c** These are good suggestions, scoring 2 marks. **d** This is correct, for 2 marks. **e** This is correct, for 2 marks. **f** Two correct points, scoring 2 marks out of 3.

Question 7

The diagram below shows a kidney nephron. Filtration occurs between the glomerulus and Bowman's capsule and, as the filtrate moves through the rest of the nephron, changes occur in the composition.

(a) Changes in the composition of the filtrate as it moves through the nephron can be compared with the composition of the blood plasma and expressed as a filtrate:plasma ratio for individual substances. The graph below shows changes in the filtrate:plasma ratio for a number of substances within the proximal tubule.

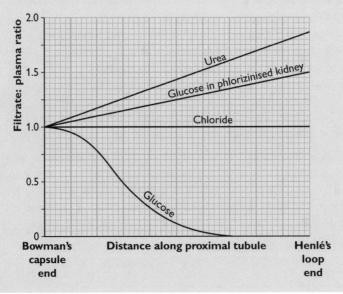

> (i) The graph includes the filtrate:plasma ratio for glucose in a kidney that was treated with phlorizin. Phlorizin inhibits glucose transport in cells. Suggest an explanation for the change in the filtrate:plasma ratio for glucose in phlorizinised kidney. **(2 marks)**
>
> (ii) Explain the results for the following substances, in an untreated kidney, taking account of your answer to (i): glucose, urea and chloride. **(5 marks)**
>
> (b) In mammals that live in deserts, the loop of Henlé is much longer than in mammals living in environments where water is plentiful. Suggest an explanation for this adaptation. **(3 marks)**
>
> **Total: 10 marks**

ⓔ The examiner has provided a diagram of a nephron to which you may refer when answering the question parts. Part (a) shows a graph that will be unfamiliar and which you must study carefully in order to answer the question sub-parts. Read the labels. The *y-axis* shows the filtrate-to-plasma ratio for different substances. A ratio of 1 means that the concentrations in the filtrate and the plasma are the same and this is the situation following filtration into the Bowman's capsule. Work out what an increase or decrease in the ratio will mean. Having understood this, you then apply your understanding of reabsorption in the proximal tubule. Part (b) further asks you to apply your understanding of the role of the loop of Henlé to suggest why desert mammals have especially long ones. The command *suggest* means that you are not expected to have knowledge of this but should be able to reason out an explanation.

Student A

(a) (i) Glucose has remained in the proximal tubule since it cannot be reabsorbed ✔ and since water is reabsorbed the concentration of glucose is increased. ✔ a

(ii) Glucose is actively transported ✔ out of the proximal tubule into the blood. ✔ Little urea is reabsorbed. ✗ Chloride is too big to diffuse ✗ or be actively transported ✗ so it remains in the proximal tubule. ✗ b

(b) A longer loop causes a greater salt gradient in the medulla, ✔ which causes greater permeability. ✗ More water reabsorption takes place and so less is lost by the mammal. ✔ c

ⓔ **6/10 marks awarded** a The answer is correct, for 2 marks. b The movement of glucose is well understood, but the rest is incorrect. 2 marks out of 5 awarded. c The first and last points are correct, but the second is obscure.

Student B

(a) (i) The filtrate glucose concentration increases as its reabsorption is inhibited. ✔ a

(ii) All glucose is reabsorbed into the blood ✔ by active transport ✔ in the proximal convoluted tubule. A large proportion of water is reabsorbed ✔ and so the concentration of urea in the tubule increases. ✔ As water is reabsorbed, chloride follows by diffusion ✔ while some is actively transported so that the concentration remains the same. ✔ b

(b) The longer loop means that a much larger salt gradient is provided in the medulla. ✔ This means that more water will move out of the collecting ducts. ✔ c

ⓔ **8/10 marks awarded a** Not being reabsorbed is on its own not a reason for the increase in glucose concentration — the result also depends on the fact that water is reabsorbed. I mark out of 2 awarded. **b** Five correct points included, for 5 marks. **c** These points are correct, but miss out on adding that a more hypertonic urine is produced with increased water conservation. 2 marks out of 3 awarded.

Question 8

Duckweed (*Lemna* species) grows on or near the surface of pond water and its growth can be measured by counting the number of fronds (leaf-like structures) that the plants produce.

The graphs below show the results of an experiment in which two species of duckweed, *L. gibba* and *L. polyrrhiza*, were grown separately and together. The plants were grown in tanks of pond water under identical conditions.

 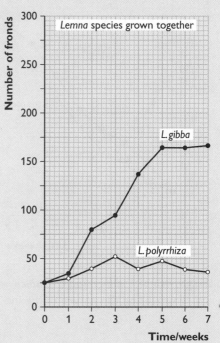

(a) With respect to the growth of *L. gibba* (grown separately) explain when intra-specific competition would be most severe. (2 marks)

(b) What is the evidence that *L. gibba* and *L. polyrrhiza* are in competition? (1 mark)

(c) Experiments on competition between two species have suggested that the species which most efficiently utilises resources, and so has the greater rate of population increase, would win out in competition.
   (i) Determine the growth rate for each species of *Lemna* when grown separately (as fronds produced per week) from 0–5 weeks. (2 marks)
   (ii) On the basis of your answer to (c) (i) above, state which of the two species might be expected to be the winner in competition. (1 mark)

(d) *L. gibba* produces special air-filled sacs in the plant body. Use this information to explain the actual outcome of competition as illustrated in the graphs above. Suggest the resource for which the plants are competing. (3 marks)

The graphs below show the results of an investigation in which two other species of duckweed, *L. minor* and *L trisulca*, were grown separately and together.

(e) Compare the growth of *L. minor* and *L. trisulca* when grown separately and when grown together, and comment on the nature of the interaction which is taking place. **(4 marks)**

Total: 13 marks

ℯ You may have met inter-specific competition graphs before (they are included in Figure 27 of this book). You can use this understanding to answer part (b). However, in part (a), notice that you are asked about *intra-specific* competition in *Lemna gibba*. In part (c) (i), you need to read values carefully from the graphs for each species to work out the growth rate. In (c) (ii), you are expected to deduce which species would out-compete the other, *on the basis of their growth rates*. You are then asked, in part (d), to *use the information provided* to explain the actual outcome of competition, *as evidenced in the graph*. Part (e) requires the interpretation of graphs showing the interaction of two other species of *Lemna*. This is entirely novel. You must practise AO2 skills so that you are confident in dealing with questions like this under examination conditions.

Student A

(a) Intra-specific competition would be most severe when the population is growing rapidly ✗ as resources become limited. ✔ a

(b) They are not reaching a high level. ✗ b

(c) (i) $(170 - 25) \div 5 = 29$ fronds week^{-1} ✔
(220 – 25) $\div 5 = 39$ fronds week^{-1} ✔ c
(ii) *L. polyrrhiza* ✔ d

(d) *L. gibba* competes better ✔ for light and blocks light from reaching the other species. ✔ **e**

(e) In normal competition, the stronger competitor still loses out when grown in mixed culture. However, *L. minor* appears to benefit from the interaction. ✔ This is not mutualism as *L. trisulca* still loses out. ✔ This must therefore be an exception to inter-specific competition. ✗ **f**

ⓔ **8/13 marks awarded a** If the population is growing then resources are not limiting. The second point is correct for the question asked. 1 mark out of 2 awarded. **b** The answer is incorrect and scores no marks. **c** This is correct, for 2 marks. **d** This is correct, for 1 mark. **e** There are two correct points here, for 2 marks out of 3. **f** Two good initial points, but the student seems to have given up in trying to provide further explanation.

Student B

(a) Intra-specific competition will be most severe at high population density after 5 weeks ✔ when the availability of resources limits the rate of growth. ✔ **a**

(b) When growing together the number of fronds produced by each is considerably less than when each species is grown separately. ✔ **b**

(c) **(i)** $170 \div 5 = 34$ fronds week^{-1}
$220 \div 5 = 44$ fronds week^{-1} ✗ ✔ **c**

(ii) *L. polyrrhiza* ✔ **d**

(d) The two plants must have been competing for light ✔ as the air-filled sacs of *Lemna gibba* allow it to float above and block light to *L. polyrrhiza*. ✔ *L. gibba* is the winner. ✔ **e**

(e) When grown separately they grow at a similar rate. However, when grown together *L. minor* grows more rapidly, ✔ while *L. trisulca* does not grow as quickly. ✔ When interacting, *L. minor* benefits from the presence of *L. trisulca*. ✔ **f**

ⓔ **11/13 marks awarded a** Both points are correct, for 2 marks. **b** This is correct and well explained, for 1 mark. **c** 1 mark is for reading the figures from the graph. However, the student has forgotten to subtract the initial density of 25 in each case, so failing to score a further mark. **d** This is correct, for 1 mark. **e** This is correct and well explained, for 3 marks. **f** There are three points worthy of marks. However, there are other points that may have been made (e.g. that this is *not* competition, it is a +/− interaction; *L. minor* is obtaining some substance from *L. trisulca* beneficial to its growth), although admittedly this is a very novel situation.

ⓔ **Section A total for Student A: 45 marks out of 72. Section A total for Student B: 62 marks out of 72.**

Section B

Question 9

Write an essay to

(a) outline how transmission occurs across synapses and along neurones, (10 marks)

and

(b) discuss how each of the following drugs interferes with neurotransmission.

Drug or poison (and origin)	Effect
ω-Agatoxin (funnel web spider)	Blocks calcium receptors
Batrachotoxin (poison dart frogs)	Depolarises membrane by opening sodium channels
Morphine (opium poppy)	Blocks receptor sites on post-synaptic membrane
Organophosphates (man-made nerve gases)	Inhibit the action of cholinesterase enzyme
Tetrodotoxin (puffer fish)	Blocks sodium channels so preventing action potentials

(6 marks)

Quality of written communication is awarded a maximum of 2 marks in this section. (2 marks)

Total: 18 marks

 You are provided with a structured essay in section B. Part (a) requires you to outline transmission, both across synapses and along neurones. You must ensure that you cover both and that answers are detailed if you are to access all the marks available. For such a long answer, you will obviously benefit from making out an essay plan beforehand. While the question part is AO1, it is not just simple recall. You need to think carefully about the sequence of actions that will take place from the moment that an impulse arrives at a synoptic knob. Part (b) is entirely AO2 and requires you to interpret the actions of five drugs or poisons. You will have to read the information carefully in order to work this out. Quality of written communication (QWC) is assessed in this question and will be based on how well you sequence your ideas and are able to draw out the links between cause and effect.

Student A

(a) Following stimulation of receptors, waves of excitation are passed along ✗ sensory neurones until they reach synaptic knobs. In the synaptic knob, Ca^{2+} ions initiate the process whereby vesicles, containing acetylcholine, fuse with the pre-synaptic membrane ✔ and pass acetylcholine via exocytosis. ✔ Acetylcholine, the transmitter substance, binds with the receptors on the post-synaptic membrane ✔ and opens up protein channels. Na^+ ions then pass through the post-synaptic membrane causing it to be depolarised. If sufficient depolarisation ✔ occurs an action potential is achieved and the signal is passed onward. Once on the motor neurone, the action potential moves along the axon ✗ to the effector, where the response takes place. Both the sensory and motor neurones are myelinated, meaning that the signal is passed on with greater velocity. This is due to the signal 'jumping' from one node of Ranvier to the next. This is called saltatory transmission. ✔ a

(b) ω-Agatoxin 'blocks calcium receptors' and so prevents synaptic transmission from being initiated, ✔ as it is Ca^{2+} that binds to receptors on the presynaptic membrane that start the response.

Batrachotoxin 'depolarises the membrane by opening sodium channels' which means that the post-synaptic membrane is constantly depolarised and the absolute refractory period is constant.

Morphine 'blocks receptor sites on the post-synaptic membrane' so that acetylcholine cannot attach and no depolarisation of the post-synaptic membrane takes place. ✔

Organophosphates 'inhibit cholinesterase', which means that acetylcholine is not broken down ✔ and cannot be passed back to the pre-synaptic side.

Tetrodotoxin 'blocks sodium channels' so Na^+ cannot pass into the post-synaptic membrane and so no action potentials can be achieved. b

QWC c

ⓔ **9/18 marks awarded** a Five appropriate points are provided. The essay lacks detail, which is virtually absent when describing the transmission of an impulse along the neurone. For example, with respect to synaptic transmission, where do the calcium ions come from? Regarding impulse transmission, there is nothing about the potential difference across the axon membrane, depolarisation of the neighbouring part of the axon membrane, recovery of resting potential or the impulse as a self-perpetuated action potential. b The student too often repeats the information in the stem without making deductions. Still there are three correct points, for 3 marks out of 6. c The student expresses ideas clearly but the phrasing and use of biological vocabulary are somewhat limited, so only 1 mark is awarded for quality of written communication.

| Student B |

(a) Transmission occurs across synapses following the arrival of an action potential at a synaptic bulb. This causes the pre-synaptic membrane to become permeable to calcium ions, which enter the synaptic bulb. ✔ This causes the synaptic vesicles to fuse with the synaptic membrane, ✔ secreting acetylcholine into the synaptic cleft. ✔ The acetylcholine diffuses across the cleft and fuses with receptor sites on the post-synaptic membrane. ✔ This causes an influx of sodium ions creating an excitatory post-synaptic potential (EPSP). ✔ If this reaches a threshold level an action potential is evoked. ✔

Neurones or axons have a resting potential when they are negative on the inside. ✔ When stimulated the potential difference is reversed and an action potential occurs. ✔ The depolarised section then undergoes repolarisation and has a refractory period ✔ when no further action potentials can occur. However, a local circuit with the neighbouring section ✔ causes depolarisation and an action potential there. ✔ The result is that action potentials are propagated down the length of the axon. ✔ a

(b) ω-Agatoxin 'blocks calcium receptors' and so no Ca^{2+} ions can enter the synaptic knob meaning that the synaptic transmitter will fail to be released. ✔

Batrachotoxin 'depolarises the membrane by opening sodium channels', which means that the impulse is unable to pass along the neurone. ✔

Morphine 'blocks receptor sites on the post-synaptic membrane' so that acetylcholine cannot bind with them and set up an excitatory post-synaptic potential. ✔

Organophosphates 'inhibit cholinesterase' meaning that the ion channels are kept open and impulses continue to be fired in the post-synaptic neurone. ✔

Tetrodotoxin 'blocks sodium channels' so that action potentials are prevented. **b**

QWC **c**

ⓔ **16/18 marks awarded** **a** The detail and balance between synaptic and impulse transmission are good and so the full 10 marks are awarded. **b** The actions of the first four drugs are well explained. However, the answer regarding tetrodotoxin is simply a restatement of the information in the question; the student needs to add that no impulse can be fired in the post-synaptic membrane. Further, the sixth mark, for reference to the ultimate effect of paralysis or loss of sensation, has not been achieved. 4 marks out of 6 awarded. **c** The student expresses ideas clearly and fluently through well-sequenced sentences. The use of biological vocabulary is good throughout. 2 marks are awarded for the quality of written communication.

ⓔ **Section B total for Student A: 9 marks out of 18. Section B total for Student B: 16 marks out of 18.**

Paper total for Student A: 54 marks out of 90, a grade C. Overall, while this is a reasonable performance, there are obvious opportunities for improvement. Revision could have been more thorough as is evidenced in the improvement that could have been made in those question parts requiring straightforward knowledge (e.g. impulse transmission along neurones in question 9 (a)). Also, an understanding of biological terms (a weakness indicated in question 1) would be improved by compiling a glossary of terms for each topic. It only takes one or two marks in each question to raise the grade significantly. There is also room for improvement when it comes to the development of skills (such as analysis and evaluation) and this is most readily achieved with greater practice of questions, particularly those that present information (in a variety of forms) in unfamiliar contexts.

Paper total for Student B: 78 marks out of 90, a grade A. This is a very high level of performance and there is ample evidence that the student has a detailed understanding of the topics in A2 Unit 1, balanced with the ability to apply that understanding in unfamiliar situations.

Knowledge check answers

1. To increase the length of the tubule and so increase the surface area for reabsorption.

2. The pressure is high because the renal arteries are short and close to the heart, and because the afferent blood vessel carrying blood to the glomerulus is wider than the efferent vessel.

3. **(a)** The endothelial cells are thin and contain pores, which allow all constituents of the blood plasma through. **(b)** The basement membrane is permeable to small molecules (smaller than those with a relative molecular mass of 68 000). **(c)** There are numerous slits between the foot-like processes of the podocytes, which allow free flow of substances into the cavity of the capsule.

4. It is the same (because the basement membrane is permeable to glucose).

5. Some of the molecules in the nephron are waste (e.g. urea) and must be left in the fluid to be excreted. Other molecules are useful to the body and must be reabsorbed.

6. **(a)** Osmoreceptors of the hypothalamus; **(b)** pituitary gland (posterior lobe)

7. More ADH

8. A short loop of Henlé results in less water being reabsorbed but because water is readily available to beavers (they live beside or in water) they do not need to conserve it so much.

9. A parasite is an organism that lives in or on its host for all or part of its life; it causes harm and gains nutrition from its host. A pathogen is any organism that causes disease.

10. The bacteria in the intestines are adapted to the conditions there and competitively exclude pathogenic bacteria.

11. Our own cells carry antigens, just like any other cell. The antigens associated with our own cells are recognised as 'self'. This is due to a process during development that destroys any cells from the immune system that have receptors complementary in shape to the antigens on our own cells. Therefore, once the immune system is mature, it contains no cells with receptors that attach to our own antigens and stimulate a response.

12. Only one or a few B lymphocytes are specific (carry the appropriate receptor) to an antigen. Each sensitised B lymphocyte then produces many memory cells, which can respond to the antigen if met again.

13. Helper T-cells help coordinate the immune response; they activate B-cells and stimulate phagocytes. Killer T-cells attack and destroy infected host cells by injecting toxins into cells that show signs of infection.

14. If the helper T-cells are destroyed by viral action, the person will have a weak immune system and will be more susceptible to infections.

15. Bottle-fed babies do not receive the antibodies contained in the mother's milk.

16. *Differences*: B-cells respond to 'free' bacteria and viruses, while T-cells respond to infected cells, cells changed by cancer or cells transplanted from another individual; B lymphocytes divide to produce plasma cells and memory cells, while T lymphocytes divide to produce killer T-cells, helper T-cells and memory T-cells. *Similarities*: both have specific surface proteins that match the antigens (on the surface of the bacterium or the infected cell); both exhibit a primary response (via the production of antibodies or the action of killer T-cells) and a secondary response (via the production of memory cells).

17. Memory lymphocytes for the disease would be present in the immune person and not in the non-immune person.

18. Each growth substance has a specific shape that will only bind to a specific receptor with a complementary shape on the surface membrane of particular cells.

19. That light results in the lateral displacement of auxin to the dark side; and does not destroy auxin.

20. The coleoptile grows towards the light if the tip is intact and light can get to it. Cutting off the tip or stopping light getting to it results in no growth towards the light.

21. Mica stops passage of chemicals and prevents the growth towards light. Gelatin allows passage of chemicals and growth towards light occurs normally.

22. Far-red light would cause the rapid conversion of P_{730} to P_{660}; the level of P_{730} would be reduced and flowering would not occur.

23. The CNS (grey matter) — motor neurones transmit impulses from the CNS to effectors.

24. Schwann cell

25. The resting potential is determined by an unequal distribution of charged ions inside and outside a neurone, making the inside negative to the outside. This is mainly due to an excess of sodium ions (Na^+) outside.

26. During the refractory period the membrane, immediately behind that exhibiting an action potential, is unexcitable and so the impulse can only be propagated forward; the refractory period is largely responsible for the unidirectional propagation of action potentials along neurones.

27. **(a)** The impulse would be slower as it would have to make more jumps. **(b)** No impulse because the nodes are too far apart to form a circuit; too far for the action potential to jump.

28. **(a)** Produces the vesicles. **(b)** Produce ATP for the synthesis of neurotransmitter substance (e.g. acetylcholine from choline and ethanoic acid diffusing back into the bulb).

29. Calcium

30. **(a)** Curare will prevent further transmission (along neurones and muscle fibres) and lead to paralysis (so that breathing stops). **(b)** Sarin will lead to the continued stimulation of neurones and muscle fibres resulting in muscle convulsions and leading to muscle fatigue (so that breathing stops).

31. The cornea refracts light as it enters the eye so that light is focused onto the retina; the cornea refracts light more than the lens though the lens adjusts the focusing for near and far objects.

32. The lens fattens (thickens).

33. Cones capable of colour perception operate in bright light; in dim conditions the cones cannot operate and the rods, which can operate in dim light, lack colour vision.

34 Blue, green and red

35 Binocular vision is simply vision with two eyes. Stereoscopic vision involves the ability to combine the images formed by the two eyes to produce a single, three-dimensional image.

36 Cardiac, smooth and skeletal muscle

37 I bands, H zones and sacromeres become shorter, while A bands stay the same.

38 (a) Calcium ions unblock the binding sites on the actin filaments by moving the blocking proteins, so that myosin heads can bind onto actin filaments, initiating the formation of acto-myosin bridges and muscle contraction. **(b)** ATP is used to break the acto-myosin bridges so that the cycle of attachment, rotation and detachment can continue.

39 (a) Almost equal, but the reproduction rate is slightly higher. **(b)** Reproduction rate is higher. **(c)** Equal.

40 Abiotic factors: light and mineral ions. Biotic factors: competition, grazing and predation.

41 During the exponential phase, intra-specific competition is minimal because resources are in plentiful supply. However, during the stationary phase, intra-specific competition is intense because resources are limited.

42 An r-selected species

43 Predators kill their prey to obtain nourishment; parasites take nutrients from their host but generally do not kill them.

44 Mutualism

45 (a) It takes time for the predator to respond to changes in food supply, for example time to reproduce in response to an increased food supply. **(b)** The predator does not eat all the available prey/energy is lost between trophic levels.

46 Competition would eventually lead to the exclusion of one species.

47 Highly species-specific, long term, no environmental contamination

48 Pioneer species, through their activities, may create a soil suitable for the next stage of succession.

49 Light

50 Primary succession involves the introduction of plants and animals into an area that has not previously supported a community. Secondary succession refers to the reintroduction of organisms into an area previously occupied but perhaps destroyed by fire. Secondary succession usually takes place faster because it occurs where soil, often containing seeds, is present.

51 Hawks occupy both the secondary and tertiary levels.

52 Because the trees are very large and represent a huge biomass.

53 (a) While water is valuable is does not have an *energy value* for the consumer. **(b)** Heating above 105°C may burn the plant material, with organic material being broken down by combustion.

54 Because the producers (first trophic level — phytoplankton) are reproducing at a sufficiently high rate to support the larger biomass of consumers (second trophic level — zooplankton), i.e. the producers have a greater productivity.

55 Because NPP represents the total energy available to consumers and the whole ecosystem.

56 NP = A − (R + U)

57 Because the NPP of the tropical forest plants is much greater than the NPP of the arctic plants and so there is much more energy available to support consumers.

58 A protein-rich diet is more readily and efficiently digested; faeces contain less undigested matter as there is no cellulose in the diet.

59 $\dfrac{13\,472}{83\,240} \times 100\% = 16.2\%$

60 Eating plants directly eliminates further links in the food chain and so the energy 'loss' (approximately 90% between one trophic level and the next) is reduced.

61 Photosynthesis/carbon dioxide fixation

62 Nitrate ions (from the soil)

63 (a) Decomposition/ammonification; **(b)** nitrogen fixation; **(c)** nitrification; **(d)** denitrification

64 Denitrifying bacteria

65 Combustion of fossil fuels and deforestation

66 (a) Higher temperatures stimulate algal growth (increasing enzyme activity and algal metabolism). **(b)** movement of the water (e.g. tumbling over rocks) re-oxygenates the water.

67 Organic matter needs to be broken down to release nutrients such as ammonia (converted to nitrates by nitrifying bacteria) and this takes place over time.

68 To develop plants that grow well with available nutrient (fertiliser) levels; to produce drought-, disease- and pest-resistant crop plants.

69 Management techniques ensure that biodiversity is maintained, while providing a continuous and regular supply of wood.

A

abiotic factors 43, 48

ABO blood system 21

accommodation, lens of
eye 35–36

acetylcholine 32–33

acid rain 63, 90

actin filament 38, 39, 40

action potentials 31, 37, 39, 40

active immunity 19–20

agglutination 21

agglutinins 17, 21

agriculture 56
intensive farming 66–67
productivity 56
sustainable farming 67–68

algal blooms 65

all-or-nothing law 31

animal-farming methods 56

antibody-mediated
immunity 16–17

antibody structure 16

antidiuretic hormone (ADH) 11, 29

antigens 15, 16, 17, 18, 19, 20
and blood transfusion 21–22

antitoxins 17, 19

apical meristem 24, 25

artificial active immunity 19, 20

artificial fertilisers 61, 65, 67,
89–90

auxins 24, 29, 98–99
action of 25
and phototropism 25–27

axons 30, 31, 32, 33

B

bacteria 14
antibody-mediated
immunity 16–17
and organic pollution 64
phagocyte engulfing 15
role in nitrogen cycle 58–59

basement membrane 9, 10

B-cells 15, 16, 17, 19

binocular vision 38

bioaccumulation 66

biodegradable 68

biological control 46–47

biological oxygen demand
(BOD) 64

biomass
and productivity 53
pyramids 52
and succession 49

biotic climax (community) 48

biotic factors 43

biotic potential 42

blind spot 35, 36

blood groups 21, 22

blood transfusion 21–22

blue-green algae 65

B lymphocytes 15–16

bone marrow 15, 16, 21

booster (of vaccine) 20

Bowman's capsule 8, 9, 104–05

C

carbon cycle 57–58

carbon dioxide 57, 61, 62, 63, 102

cardiac muscle 38

carnivores 55

carrying capacity 42

cataracts 63

cell enlargement 24–25

cell-mediated immunity 17–18

central nervous system
(CNS) 29–30

centron, neurone 30

chlorofluorocarbons (CFCs) 61, 62

ciliary muscles 35

circular muscles 35

climatic climax (community) 48

climax community 48

clumping 21

coleoptiles 26, 98–99

collecting ducts 8, 11

colour vision 37

communities 47–50

competition, inter-specific 44,
45–46

competitive exclusion principle 46

cones 36, 37, 77–78

consumers 51, 52, 54, 55

contraction of muscle 39–40

coordination
in flowering plants 23–29
in mammals 29–40

cornea 34, 35

cortex, kidney 8

cranial nerves 30

crop rotation 66

cytokinins 24

D

dark adaptation 37

decline phase, population
growth 42

decomposers 51
and energy transfer 54, 55, 56
nutrient cycling 57, 58

dendrites 30, 33

dendrons 30, 31

depolarisation, neurones 31,
32, 33

depth and distance perception 38

detritus food chain 50

diseases, infectious 13–14

distal convoluted tubule 8, 10

driving force, ultrafiltration 9

E

ecological energetics 50–57

ecological niche 50

ecological pyramids 51–53

ecological succession 48–49

economic damage threshold 46

ecosystems
  communities 47–50
  ecological energetics 50–57
  human impact 60–69
  inefficiency of energy
    transfer 53–56
  nutrient cycling 57–59
  populations 41–47
effectors 29, 30, 34
efferent arteriole 8, 9
egestion 7
endotherms 55
energy budgets 54, 75
energy/productivity, pyramid
  of 52–53
energy transfer in
  ecosystems 53–56
enhanced greenhouse effect 61
environmental resistance 42
environment, human impact
  on 60–69
ethical issues, farming 56
eutrophication 65, 67
exam skills 6, 70
excretion 7–10
exponential (log) phase,
  population growth 42
eye 34–38
  light reception 35–38
  questions & answers 77–78,
    100–01
  structure and function 34–35

F
farming see agriculture
fertilisers 56, 65, 67, 89–90
filaments, muscle 39, 40
filter, kidney 9
filtrate, composition of 10
filtration see ultrafiltration
flowering plants 23–24
  auxins and phototropism
    25–27

phytochromes and control of
  flowering 27–28
plant growth 24–25
questions & answers 71–72
versus animal
  coordination 28–29
focusing of light 35–36
food chains/webs 50–51
food production see agriculture
foreign (non-self) antigens 15
forestry, sustainable 68–69
fossil fuels 57, 61

G
generator potential 37
gibberellins 24
global warming 60–62
glomerulus 8, 9
  net filtration pressure 12
GPP (gross primary
  productivity) 53
grazing food chain 50
greenhouse effect 60–61
greenhouse gases 61
gross primary productivity (GPP) 53

H
helper T-cells 17, 18
herbicides 66
herbivores 51, 55, 75
homeostasis 7
  kidney and excretion 7–10
  osmoregulation 11–12
hormones 11, 29
human impact on
  environment 60–69
hydrostatic pressure,
  ultrafiltration 9, 12
hyphae 45
hypothalamus 11

I
immune response 15

immunity 13–14
  active and passive 18–20
  barriers to pathogen entry 14
  and blood transfusion 21–22
  and organ transplant 20–21
  phagocyte activity 15
  specific immune
    response 15–18
immunosuppression 21
impulse propagation 32
'indicator species' 64
inefficiency of energy
  transfer 53–56
infection, pathogens
  causing 14
infoldings 10
inorganic (artificial) fertilisers 61,
  65, 67, 89–90
insecticides 66
integrated pest management
  (IPM) 68
intensive farming 66–67
internode 24
inter-specific competition 45–46
intra-specific competition 41, 43
intrinsic rate of natural
  increase 42, 43
iodopsin 37
iris 34
  control of pupil diameter 35

K
kidneys 7–10
killer (cytotoxic) T-cells 17, 18, 19,
  23, 94–95
K-selected species 43–44
K-strategy/strategist 43, 44

L
lag phase, population
  growth 42
lens 34, 35
  accommodation of light 35–36

light
  human eye's response to 35–37
  plant response to 23–28
local circuits, impulse
    propagation 32
long-day plants (LDPs) 27–28
loop of Henlé 8, 10
lymphocytes 15–16
lysozyme 14

M

macrophages 15
mammalian coordination 29–30
  the eye 30–38
  muscles 38–40
  neurones and synapses 30–34
medulla, kidney 8, 10
memory cells 17, 19, 20
memory T-cells 18
methane (CH_4) 61
microvilli 10
mitochondria 10
monocultures 66
motor neurone 30
mucus 14
muscle
  contraction 39–40, 41
  questions & answers 73–74
  structure of skeletal 38–39
mutualism 44–45
myelin sheath 30, 31, 32, 33
myofibrils 38, 39, 40, 73
myosin 38
myosin heads 39–40

N

natural active immunity 19, 20
natural passive immunity 19, 20
negative feedback 11
nephrons, kidney 8
nerve impulses 30–32
nerves 30

nervous system 29–30
net filtration force 12
net primary productivity
    (NPP) 53
net secondary productivity
    (NP) 54
neurones 30–34
neurotransmitters 32–33
niches 45, 46, 48, 49, 50
nitric oxide (NO) 63
nitrifying bacteria 58
nitrogen cycle 58–59, 79
nitrogen dioxide (NO_2) 63
nitrogen-fixing bacteria 58
nitrous oxide (N_2O) 61
nodes of Ranvier 31
non-biodegradable 66
non-renewable resources 42
noradrenaline 33
NPP (net primary productivity) 53
numbers, pyramid of 52
nutrient cycling 57–59

O

optic nerve 34, 35, 36, 37
organic fertilisers 67
organic pollution of
    waterways 63–65
organ transplant 20–21
osmoreceptors 11
osmoregulation 7, 11–12
ozone depletion 62–63
ozone layer 62
ozone (O_3) 61

P

P_{660} and P_{730} 27
passive immunity 18–19
pathogens 13–14
  barriers to entry 14
  immune response to 15–20
pelvis, kidney 8

pesticides 56, 66
  biological control 47
  questions & answers 89–90
pests 46
  integrated pest management 68
phagocyte activity 15
photoperiods 27–28
photosynthesis 57
phototropism 25–27
  questions & answers 98–99
phytochrome pigments 27
phytochromes and control of
    flowering 27–28
pigments
  in flowering plants 27
  in rods and cones 37
pioneer plants 48
pituitary gland, posterior
    lobe 11
plant growth substances 24–25
  auxins and phototropism 25–27
  phytochromes and control of
      flowering 27–28
plasma cells 16, 17
pollution
  via the atmosphere 60–63
  of waterways 63–65
polyculture 67
polymorphs 15
population 41
population dynamics 43–44
population growth
  equation 43
  factors influencing 43
  phases of 42
  questions & answers 83–84
population interactions 44–47,
    84–85
positive phototropism 25, 26–27
predator–prey interactions 45
primary consumers 51, 52, 54, 55
primary productivity 53

primary response 19, 20
primary succession 48–49
producers 51, 54, 55
productivity 53–56
 pyramid of 52–53
protein receptors 15
proximal convoluted tubule 8, 10
pupil diameter, changes in 35
pyramid relationships 51–53

R
radial muscles 35
reabsorption 8, 10, 12
 questions & answers 87, 105
receptors 28
 osmoreceptors 11
 photoreceptors 36–37
 protein 15
red blood cells, antigens 21, 22
refractory period 31
renewable resources 42
resistance, environmental 42
resources 41
 renewable and non-renewable 42
respiration 53, 54, 55, 56, 57, 58
resting potential 31
retina 34, 35
 questions & answers 77–78
 rods and cones 37
 structure of 36
retinal convergence 36, 37
Rhesus system 22
rhodopsin 37
rods 36, 37, 77–78
r-selected species 43–44
r-strategy/r-strategists 43–44
run-off 65

S
saltatory conduction 32
sarcolemma 38, 39
sarcoplasmic reticulum 38, 39
Schwann cells 31

secondary consumers 51, 52, 53, 54, 55
secondary productivity 53, 54
secondary response 19, 20
secondary succession 49
sensory neurone 30
seres (stages) 48, 49
short-day plants (SDPs) 27–28
silage 56
skeletal muscle 38–39
skin cancers 63
sliding filament theory 39–40
slurry 63, 67
smooth muscle 38
soil erosion 67, 68
solar energy 53
specific immune response 15–20
spinal nerves 30
stationary phase, population growth 42
stereoscopic vision 38
study skills 5–6
succession 48–49
sulfur dioxide (SO_2) 63
summation 37
sustainable farming 67–68, 89–90
sustainable forestry 68–69
synaptic bulbs 30
synaptic transmission 32–34

T
T-cells 15, 16
temperature
 increase in 61–62
 and population growth 43
tertiary consumers 51, 52, 53, 54, 55
threshold value, potential difference 31
thymus gland 16
tissue typing 20–21
T lymphocytes 15–16, 94–95

transplants, organ 20–21
trichromatic theory of colour vision 37
trophic levels 51
 and inefficiency of energy transfer 53–56
 pyramid relationships 51–53
T-tubules, muscle 38, 39

U
ultrafiltration 8, 9–10, 12
 questions & answers 86–88, 104–05
ultraviolet radiation 62, 63
universal donors 22
universal recipients 22
urea 7
urinary system 7–8
urine production 8–10

V
vaccination 20
vegetarian diet 56
viruses 13, 14
 antibody-mediated immunity 17
 cell-mediated immunity against 18
 questions & answers 81–82, 94–95
visual acuity 37
visual fields 38

W
water potential of blood, regulation of 11–12
waterways, pollution of 63–65
woodland, sustainable management of 68–69

X
X-rays 21

Z
zone of elongation 24, 25